Instruction Modeling

Instruction Modeling

Developing and Implementing Blended Learning Programs

GEORGE A. KHACHATRYAN

OXFORD
UNIVERSITY PRESS

OXFORD
UNIVERSITY PRESS

Oxford University Press is a department of the University of Oxford. It furthers the University's objective of excellence in research, scholarship, and education by publishing worldwide. Oxford is a registered trade mark of Oxford University Press in the UK and certain other countries.

Published in the United States of America by Oxford University Press
198 Madison Avenue, New York, NY 10016, United States of America.

CIP data is on file at the Library of Congress
ISBN 978–0–19–091070–9

9 8 7 6 5 4 3 2 1

Printed by LSC Communications, United States of America

For my father

Contents

Foreword

It is my great pleasure and honor to introduce a book that will share with a broad and nontechnical audience one of the most important "behind the scenes" developments in education in recent years: the method of instruction modeling and the Reasoning Mind family of blended learning systems for education that it has produced.

Instruction modeling, as Dr. Khachatryan defines it, is the process of studying the experiences of students in world-class mathematics classes, building online lessons that replicate some of the core parts of these experiences, and training and supporting classroom teachers in providing the other core parts of these experiences.

In this definition, Dr. Khachatryan captures several key aspects of what good educational design is and should be. The first is that it should build off of the best of what we already know. Around the world, teachers and curriculum developers strive every day to improve education. There's no reason to start from scratch. We should find what works extremely well and find out how to replicate it. This is what Dr. Khachatryan and his colleagues have done, searching out the best schools in a country with an impressive mathematical tradition, and building off of the years of careful curricular development there. At the same time, they have researched and learned from the best work in educational interface design.

The second is that Dr. Khachatryan avoids the tendency to overemphasize the role of either technology or teachers. Creating great learning experiences is *difficult*. By building off what both computers and teachers are good at, we have a much better chance at providing truly excellent education than is feasible with either alone. Despite the evidence going back decades that teachers hold the key to the success of computer-based learning—see, for instance, Janet Schofield's brilliant book *Computers and Classroom Culture*—far too many computer-based curricula ignore or even devalue the teacher. At the same time, no teacher can provide feedback to 25 students at once, and practices such as the knowledge tracing discussed in this book are essentially impossible for a teacher to realize for an entire classroom at once (despite attempts to do so for decades, for example in Benjamin Bloom's attempts to implement mastery learning in American classrooms in the 1970s).

Computers can provide explanations and immediate feedback on problems very well; teachers excel at supporting struggling students. By leveraging both, and using technology to support and empower teachers rather than to reduce them to classroom monitors, Reasoning Mind approaches the "best of both worlds" potential that blended learning promises to someday offer. Clearly, there is much work still to come to fully reach that potential. Only a small number of systems and curricula have started down that path, and this book makes the case that instruction modeling is a key part of it.

The result of the intense effort conducted by Dr. Khachatryan and his colleagues (and one must admire the sincerity of Dr. Khachatryan's promise of "generations of effort" in a "laborious approach"; but as he says, there really is "no shortcut" to amazing education) is a set of curricula that have been repeatedly shown to be educationally effective. Chapter 6 details the 15 studies (so far) that demonstrate various aspects of Reasoning Mind's effectiveness as well as where and when it has been less successful. Only a small number of other systems, such as Cognitive Tutor, ASSISTments, and ALEKS, even approach having the same degree of evidentiary basis.

More importantly, Reasoning Mind is not just a system that has been thoroughly studied through academic research. It also works at scale, improving the lives of hundreds of thousands of children.

Included in that number are my own children. When my daughter was struggling with mathematics in elementary school (using a much inferior blended learning system in her class, one whose selection was justified by the school district's technology specialist based—by his own admission—solely on the evidence on the company's website), I knew from my research into blended and online learning environments (having formally researched and published on more than twenty) that Reasoning Mind's *Foundations* course was the right choice for her. I obtained a copy for her—she caught up to where she should be (and continues to work on mathematics considerably ahead of grade level at home). Her brother, starting with Reasoning Mind at home from the very start, quickly sped past grade level and continues to excel with mathematics to this day. Online characters like "Higgie, the giant hamster" were welcome friends that helped illustrate difficult mathematics to them.

An anecdotal report is no substitute for real data, of course (of which there is a considerable amount). But hopefully it helps to illustrate that the approach that Dr. Khachatryan introduces here does not just lead to better test scores (as many systems do). It leads to a deep and sustaining understanding of and enthusiasm for mathematics. It changes lives.

This book provides key detail—an "insider view," if you will—as to what is necessary to develop exemplary learning experiences for students. In a world where I have heard some curriculum developers pride themselves on writing journal papers and books that give insufficient detail to a reader to replicate their approach, this book stands out for its attention to the key details that are essential for a blended learning system like Reasoning Mind to work. This openness makes it possible to build off past efforts and speeds the development of future innovations; this book is not just about how one system succeeded, but about practices that future developers can take to develop systems that succeed. In one of the seminar courses I teach at the University of Pennsylvania, I devote a class session to discussing the Reasoning Mind model for implementation and its implications both for the developers of future learning systems and for practice at schools. This book provides the type of detail that will make it feasible for a much broader range of practitioners and researchers to use these methods. Instruction modeling is an approach with many possible uses. Curriculum developers can use instruction modeling to develop learning systems that emulate the practices chosen by exemplary teachers. Education researchers can use instruction modeling to better understand what practices work, and in what situations. Teachers can leverage these efforts to learn what practices are best, and to become exemplary teachers themselves. This is especially feasible when those teachers are using a curriculum that supports the development and use of these practices.

In this pivotal moment in education, we are undergoing technical shifts that are transforming the practice of education. But we must be careful to promote the practices that are best for students, not simply those practices that become easiest. Instruction modeling has the potential to be a key method in our research and development toolbox, an essential tool for enhancing the quality of learning for all learners.

Prof. Ryan S. Baker
University of Pennsylvania Graduate School of Education
May 9, 2018

Instruction Modeling

Introduction
Grandmothers and Machines

THIS STORY BEGINS IN 1989, WHEN MY PARENTS LEFT the Soviet Union and came to America. It was a gutsy move: if their request to leave was rejected by the Soviet government, they would be fired from their jobs and permanently barred from emigrating. Not to mention, my parents knew little about life in the United States. Some of their friends had left a decade earlier, but contact with foreigners was restricted. They had a general sense that America was a wonderful place to be, but beyond this did not know much.

Per the family folklore, the decision to leave was made when my father returned from a three-week work trip. During his absence, my mother had been reading recently published books chronicling the regime's depredations. Everyone knew that they lived in an oppressive society, but until Mikhail Gorbachev led reforms that allowed the free publication of books that told the true story, few realized the extent. My great-grandfather had endured ten years of hard labor in Stalin's Gulag, but he—and most like him—never talked about it. The Soviet Union of the 1980s was in many ways a milder place

Instruction Modeling. George A. Khachatryan, Oxford University Press (2020)

DOI: 10.1093/oso/9780190910709.001.0001

than it had been a generation earlier, but the now-published histories underscored the fact that, milder or not, the entire system was built to limit and control its citizens. When my father returned from his trip, and before he had time to set down his bags, my mother told him: "We're leaving."

My parents were granted permission to emigrate. We received refugee status and landed at John F. Kennedy International Airport in New York. I still remember the car ride from the airport: a five-year-old child, I was given a balloon and some chocolates. That was an auspicious start. We were received by the Jewish community in Monroe, a small town upstate. They supported us while my father searched for work.

Eventually, my father was offered a position as an assistant professor, working on artificial intelligence at Texas A&M University in College Station. My parents had heard about Texas only that there were cowboys, and that it was very hot. My mother enrolled in English courses at the university.

Just a year later, the Soviet Union collapsed. My grandparents—whom my parents had feared they would never see again—were able to join us in Texas. Among them was Victoria Khachatryan, a lifelong mathematics teacher. After attending the Moscow Pedagogical University, my grandmother had taught mathematics for thirty years. Though as a grandmother she was nothing but indulgent, I knew from my father that her students had considered her strict. In America, my grandmother made ends meet by babysitting and doing household chores. She satisfied her love of teaching mathematics by volunteering as a tutor at the local middle school.

My grandmother was not the only lover of mathematics in the family. My father was an applied mathematician. My mother was an engineer, and her father had been a college physics professor. My great-grandfather who had spent time in the Gulag had been a professor of mechanical engineering, and during his imprisonment had somehow managed to write a geometry textbook. It was never published, but we have preserved the manuscript as a family memento. My parents hoped that I would follow in the family's footsteps, becoming a mathematician or at least a physicist. But this appeared unlikely: my school mathematics classes had taught me little of the subject, and had certainly failed to provoke any interest in it. My parents moved me from school to school, trying out public and private schools of all varieties and sizes, but it was no use. I eventually settled at a school that had a wonderful program in the humanities, but the usual uninspiring fare when it came to mathematics. All of this was very surprising to my parents. They had come to the United States in the usual expectation that everything here would be better. Their expectation had been met—in everything but mathematics education.

The situation came to a head when my grandmother decided to assess my knowledge of the subject. I had just completed and passed an Algebra I course, becoming thereby a certified knower of algebra. She gave me two fractions to add. To her shock, I added the numerators and added the denominators. In all her years of teaching, not even her worst student would have arrived at the end of an algebra course so innocent of basic arithmetic. It was because of how I had been taught over the years: instead of genuinely understanding mathematics, I had been made to memorize routine procedures. The resulting knowledge—if it could be called knowledge at all—was short-lived. Most likely, at some point in the fifth or sixth grade I *had* known how to add fractions; but as I had learned it as a sequence of arbitrary steps, devoid of meaning, it was quickly and decisively forgotten. Fortunately for me, my grandmother decided to remedy the problem. She tutored me, helping me understand basic arithmetic properly. This foundation made it possible for me to successfully take more advanced mathematics classes.

A second stroke of luck came in high school. I had participated—with no particular distinction—in a mathematics competition my freshman year, and this had landed me on the mailing list for Canada/USA Mathcamp, a five-week residential summer program. We received a brochure, and my parents encouraged me to apply. While I was not particularly enamored of mathematics, I had heard my parents talk about how wonderful it could be. I agreed to give it a go.

Mathcamp was beyond anything I could have imagined. We "campers" lived at a university, left to our own devices as if we were college students. The classes, which were all optional, were taught by enthusiastic graduate students. The material was never dull: the most exciting parts of mathematics had been extracted and made accessible to high school students. Occasional luminaries dropped in to give guest lectures and chat with the students. The instructors got to know the students personally, immersing us not only in the content of mathematics, but also in the culture of the community of mathematicians: the sorts of jokes mathematicians make, the pastimes they prefer, the values they share. It was the best possible kind of brainwashing, brilliantly done. I returned home with my mind made up: I would become a mathematician.

This happy ending would not have been possible without the two great strokes of luck I encountered, and my parents knew it. How many students had grandmothers who were extraordinary mathematics teachers? And among this already miniscule group, how many would ever have the opportunity to attend a program like Mathcamp?

This became a frequent topic of discussion in my family. What would it take to improve mathematics education in our adoptive country? If understanding and appreciating mathematics was to become the norm for all students, the only solution was to change the way it is taught in our nation's schools.

At the time, we did not realize just how difficult any kind of instructional reform was. But we had an inkling that it was not so easy as most imagine: simply changing the textbooks or mandating that teachers use a certain approach would not be enough. My grandmother, for example, was a good mathematics teacher not only by virtue of her personal qualities. She had been rigorously prepared, the alumna of a system that produced many thousands of outstanding mathematics instructors like herself. First, she had been the beneficiary of excellent mathematics education in school. Then, she had taken five years of courses—most of them focused on the substance of mathematics instruction—at the Moscow Pedagogical University. And following this, she had learned alongside her more experienced colleagues in schools. Expecting teachers who had no such preparation to know as much as she did was unreasonable, no matter how dedicated or intelligent they might be.

In other words, the challenge in instructional reform lies in *knowledge transfer*: excellent mathematics education relies on an enormous body of professional teaching knowledge. How can this knowledge be transferred? Imagine if we could send all of our nation's mathematics teachers to spend a decade in Russia, Singapore, Japan, and other countries with exceptional mathematics education, studying these countries' teaching methods to absorb everything they have to offer.

Even if we could find funding for this unusual endeavor—and the cost would rival the Pentagon budget—it would still be insufficient. The professional knowledge of teachers is only the most visible component of a strong mathematics education system; for this knowledge to persist, a country needs excellent textbooks, outstanding university programs to prepare new teachers, and plentiful opportunities for teachers to continue their professional development throughout their careers. This is what makes instructional improvement on a national scale so challenging: instead of blaming teachers, societies that wish to improve their mathematics education must ensure teachers have a whole host of requisite supports, each one requiring massive reforms.

This challenge gave my father an idea. He had worked with a form of artificial intelligence called *expert systems*. Expert systems existed exactly for this purpose: to extract the knowledge of human experts in different fields, so that this knowledge could be made available when the experts were in scarce supply.[1] For example, a computer scientist wishing to create an expert system for medical diagnostics would start by interviewing expert diagnosticians. She would ask them questions about how they reason, teasing out the mental rules they use: "If the patient has a fever, then ask about diarrhea." "If the patient has a stomach pain but no fever, then check for a hernia." With all of the rules in hand—most likely, a collection of many thousands—the computer scientist would proceed to write a program that would run the rules, just as the diagnosticians would do in their minds. The resulting program could then be used to compensate for doctor shortages, to train medical students, or to provide preliminary online diagnoses to patients prior to their hospital visits.

My father's idea was to apply this approach to education. A computer program that carried the knowledge of the world's best mathematics teachers could be used to help teach students, train teachers, and generally make mathematics education expertise more broadly available. He came to me and my mother—at the time I was a high school student—and suggested it as a family project. From the beginning my parents wanted to organize the work as a nonprofit organization. In 2000, they brought the idea to Ernie Cockrell, a respected Houston businessman and philanthropist with whom they had become friends after working together some years earlier. Ernie studied the approach to determine its viability, and eventually decided to give it a shot. He became the founding chairman of the board of trustees, and Reasoning Mind was born.

Following the ideas of expert systems, Reasoning Mind's approach is *instruction modeling*: studying the instruction provided in exemplary mathematics classrooms, and then using technology to make this instruction available to many more students and teachers. The word "modeling" indicates the aim, which is not to create something new, but, rather, to faithfully reproduce something that already exists but is so far only available to a select few.

Naively, one may think that it would be enough to just build a computer system that reasons like an expert teacher. However, education is fundamentally human: there are parts of it that simply cannot be automated. This includes the social aspects of learning, like the bond between teacher and student, the role of encouragement, and the rest. But it also includes some instructional aspects. For example, in good mathematics education, students do not simply practice routine problems or answer straightforward questions: they explain their reasoning, participate in mathematical discussions, and receive coaching on their mathematical writing. A computer can check a numerical answer for correctness, but it cannot judge the quality of an explanation or participate in a true conversation.

For this reason, instruction modeling cannot be confined to computers alone—the classroom teacher plays an important role.* In instruction modeling, those aspects of education that can be automated are provided through the online component, while those that are essentially human are provided by the classroom teacher. So, instruction modeling is a process of three steps:

Step 1. Study the experiences of students in world-class mathematics classes.
Step 2. Build online lessons that automate these experiences, replicating them as closely as possible.
Step 3. For experiences that cannot be automated, train and support classroom teachers to provide them to students.

* Put another way, instruction modeling is used for *blended learning*, which is any instruction that combines computer and human teaching.

Reasoning Mind's approach was to use instruction modeling to create blended learning math programs for elementary and middle school. Upon entering a Reasoning Mind classroom, a visitor will usually see students taking online lessons at computers. These lessons will simulate much of what takes place in exemplary mathematics classrooms: clear, mathematically correct explanations; challenging and varied problems; an appeal to genuine understanding. Meanwhile, the teacher will spend some of her time on classroom management, which includes supervising students to make sure they are on task, offering encouragement, and administering rewards (Figure I.1). The rest of the time—and in the best classrooms, this makes up the vast bulk of the teacher's activity—the teacher will provide the sort of in-person mathematics instruction that no computer can offer. She will review students' written work, teaching them to write clear, correct mathematical prose. She will use online data generated by the program to identify specific topics with which each student is struggling, providing one-on-one tutoring to correct students' misconceptions. And she will lead small-group sessions in which students will explain

FIGURE I.1 In this Reasoning Mind classroom at IDEA Public Schools, students work on computers while the teacher both manages the class and works with individuals and small groups.

their reasoning, critique each other's work, and learn how to approach nonstandard mathematical problems.

In theory, the combined effect of the online and in-person lessons is to provide the same quality of mathematics instruction that students would have received if they were to be magically transplanted to one of the world's best mathematics classrooms. It is like bringing my grandmother—or a similar master mathematics teacher—to every student. Of course, this is an ambitious goal, and developing the methods to faithfully model the work of outstanding mathematics teachers is a large undertaking. With each successive program, Reasoning Mind has stepped closer to the goal.

Instruction modeling may be appealing in theory, but does it work in practice? This is taken up in detail in chapter 6, where we discuss the body of evidence—including evaluation studies—concerning the effects of Reasoning Mind's programs. As a whole, the evidence indicates that these instruction modeling programs provide higher-quality instruction, improve students' knowledge of mathematics, and meet the needs of teachers and administrators.

What sets instruction modeling apart from most other approaches to designing blended learning programs? In my view, it has three major virtues.

1. It is by necessity a *comprehensive* approach: the instruction modeler seeks to recreate *everything* present in excellent offline instruction, not just some individual aspects that seem appealing. This is important because good education is always multifaceted, and no magic feature will by itself lead to good instruction. As any good teacher knows, successful instruction must account for a massive array of factors. Instruction modeling—unlike the fleeting fads that are so pervasive in education technology—takes this into account.
2. Instruction modeling is a fundamentally *cautious* approach: it does not aim to deliver the best imaginable education, but merely to recreate the best of what already exists. This is important because good instructional approaches have always and everywhere been the product of decades of trial and error; attempting to create one from scratch is a very long shot indeed. Instruction modeling takes the guesswork out of good instructional design.

3. Instruction modeling offers a way to incorporate the vast amount of teaching knowledge that is only obtained through *trial and error*. Excellent instructional traditions have developed strong lessons through decades of iterative improvement. They have learned which explanations confuse students, which examples are the clearest, which concepts are most easily forgotten and require frequent review, and many more such things that can only be found out through experience. Instruction modeling offers a way to access this knowledge.

In this book, I aim to share the lessons that I, my colleagues, and our collaborators have learned in sixteen years of work. Some of these lessons are specific to instruction modeling, and certainly the reader will leave the book with a good understanding of instruction modeling's purpose, value, difficulties, and methods. I hope that this information will compel other developers of educational programs to take up the approach, to the benefit of students and teachers. But other lessons are of a more general nature, and I believe that anyone with an interest in education technology—or, for that matter, education in general—would find them relevant. I once had naive views about what makes for good instruction, how to determine whether an approach is working, what factors truly matter for any program to succeed, how to implement programs in schools, how successful programs spread, and many other such matters. I hope that, by offering the opportunity to learn from my blunders and experiences, this book may in a few hours give the reader the knowledge that it took me sixteen years, with much trial and abundant error, to acquire.

The chapters fall into three groups. The first two chapters cover *prerequisite* knowledge: how to identify good instruction in the first place, and what it looks like in the case of mathematics. It is shocking how misunderstood these matters are, and these misunderstandings are at the heart of much educational confusion. Clear answers will form a foundation for all of the discussions that follow in the remaining chapters. I believe readers may also find the material informative in its own right.

Chapters 3, 4, and 5 present the core concept of instruction modeling. Chapter 3 tells the *story* of instruction modeling: how the idea came to be, how a prototype program was built and tested, how subsequent programs were designed, how the approach gradually matured and took full form. Each step carried surprises and taught us valuable lessons—sometimes painful, other times very amusing. Chapters 4 and 5 discuss two central questions related to the concept of instruction modeling: how it interacts with learning science and how programs developed using instruction modeling can be successfully implemented in real classrooms.

Chapters 6 and 7 place instruction modeling in a *broader context*. Chapter 6 surveys the evidence to answer the question "Of what use instruction modeling?" Chapter 7 compares instruction modeling to other approaches for designing blended learning programs, both arguing for its distinctive value and explaining how it can be constructively combined with these other approaches.

Chapter 8 offers a practical guide to conducting instruction modeling. It discusses the central steps, possible variations, and common pitfalls.

Chapter 9 discusses the future of instruction modeling: the new technologies that will strengthen it, the new areas in which it will be applied, and the government policies that can support its successful use.

Where are Reasoning Mind's program's today? In 2018, Reasoning Mind became a part of Imagine Learning, a developer of digital language, literacy, and mathematics programs. Through Imagine Learning's partnerships with school districts, the programs will become available to many more students and teachers. Of course, my fervent hope is that instruction modeling will be used by an ever-increasing number of educational publishers. As I argue in this book, that would be very much to the benefit of the teachers and students who use instructional technology programs.

The work described in this book was carried out over many years by over a hundred colleagues at Reasoning Mind. Each member of the Reasoning Mind team contributed in some way—whether direct or indirect—to the development and improvement of instruction modeling. While unfortunately I cannot list all of these colleagues here, instruction modeling is especially indebted to the efforts of Alex Khachatryan, Julia Khachatryan, Andrey Romashov, Steven Gaudino, Yuri Belikov, Konstantin Guarian, Victor Kostyuk, Will McGuinness, Kevin Wilson, Nadezhda Yakunina, Nathaniel Rounds, and Michael Von Korff. The work would have been impossible without the contributions of generous philanthropists, many of whom supported the work long before it had achieved broad recognition; they include Ernie Cockrell, Forrest Hoglund, Gerald McElvy, and Kevin Alexander, as well as the Bill & Melinda Gates Foundation, the Houston Endowment, the Brown Foundation, the Ewing Halsell Foundation, the Lyda Hill Foundation, Deloitte, ExxonMobil, Highland Capital Management, and hundreds of other philanthropic foundations and individual donors. An important role was also played by research collaborators, including Will Weber (University of Houston), Bob Houston (University of Houston), Hersh Waxman (Texas A&M University), Ryan Baker (University of Pennsylvania), Jaclyn Ocumpaugh (University of Pennsylvania), Xiangen Hu (University of Memphis), and Jeremy Roschelle (Digital Promise). Thousands of teachers and administrators have used Reasoning Mind's instruction modeling programs over the years, and many of them provided ideas and recommendations for improvements; those to whom I am particularly grateful include Sandy Nobles, Anne Mechler, Tom Torkelson, Pablo Mejia, Erik Twist, Robert Jackson, Roberto Martinez, and Austin Deakins.

There are also many people who helped in the preparation of this book. I thank James Talarico, Victor Kostyuk, Steven Gaudino, Ryan Baker, Michael Von Korff, and Alex Khachatryan for their detailed comments on the manuscript, which improved it greatly. I also thank Sarah Harrington, my editor at Oxford University Press, and the anonymous peer reviewers for their helpful suggestions. Finally, I am grateful to my wife, Marcela de Souza Santos, who not only provided excellent advice, but also laughed at my jokes.

Chapter 1

Identifying Good Instruction . . . *Do We Know Anything?*

THE CENTRAL IDEA OF INSTRUCTION MODELING IS TO study high-quality offline instruction, and then to use computers as an aid in recreating this instruction. But of course, for any such approach to be successful, we must be able to identify high-quality offline instruction in the first place! For this reason, any discussion of instruction modeling must begin with this question: what, exactly, is good instruction?

It may seem that the answer is obvious. Indeed, we have all had good teachers as well as bad ones, and have had little trouble telling the two apart; have learned from excellent textbooks, and also from lamentable ones; have taken courses we cherish to this day, and also ones we found tiresome and irrelevant. And yet, the question of what makes for good instruction—both in *what* to teach and *how* to best teach it—has been as controversial as any question in history. The most influential philosophers and scholars of education have disagreed bitterly.

Instruction Modeling. George A. Khachatryan, Oxford University Press (2020)

DOI: 10.1093/oso/9780190910709.001.0001

Some have argued that education should above all prepare students for their chosen careers, and others that all students should receive the same education. Some have argued that education must aim to build character, and others that it was pointless to expect anything but the "removal of ignorance."[1] Some have argued that students should be explicitly taught facts and procedures, and others that students should be gently guided to uncover them on their own. And these examples are only the beginning: a thousand controversies, great and small, have marked the history of education. And the same controversies have resurfaced in each age, dressed in different clothing but identical in essence.

The introduction of the scientific method in education—a major development of the last century—was meant to correct this state of affairs. Long before, the sciences had suffered from a similar malady, the same arguments repeated without resolution by each successive generation; by rejecting flights of philosophical fancy, and replacing them with the careful measurement and experimental testing of claims, the sciences had unlocked the possibility of forward progress of knowledge. Education researchers hope to accomplish the same. Just as the astronomer trains the telescope on distant stars, gathering data to test physical theories, so too the education researcher deploys the instruments of educational measurement—tests, surveys, observations—in search of objective educational truth.

Has this resolved the controversy? In this chapter, we will review the ability of existing methods to distinguish between good and bad instruction.

Broadly, I will argue here that education is an inherently murkier discipline than the so-called "hard sciences," such as physics and chemistry. I have found that many individuals involved in education—as teachers, policymakers, program developers, or even researchers—seem to believe that educational truth can be easily established through the tools of quantitative science: one has only to look at test scores, and they will reveal which students learn more, which curricula are most effective, and which teachers are the best. As a trained mathematician, I am on too intimate terms with the quantitative world to believe that a statistic is correct merely by virtue of being a statistic. We must look deeper and ask: what do these numbers *really* tell us?

Let me be clear: I am not against quantitative methods in education. Far from it. I believe these methods have allowed for previously unparalleled clarity on many educational questions. I have myself been involved in the design and execution of several experiments, and have been the coauthor of scientific publications in the field. What I *do* oppose is a simplistic view of the precision and sufficiency of these methods. I will argue in this chapter that quantitative methods deserve their place of honor in education, but we must nevertheless take the numbers that come from educational experiments—no matter how many places after the decimal point they boast—with a grain of salt. Most importantly, we must supplement the evidence gleaned from these experiments with insight from *humanistic* scholarship: the works of education historians, ethnographers, and critics. This multidisciplinary approach will allow us to reach far sounder conclusions than the alternative.

A note on terminology: throughout this book, I will use the word "quantitative" (or, in some cases, "experimental") to denote scholarship that employs the traditional scientific method used in the physical sciences, with its use of experiments and quantified outcomes.* I will contrast it with "humanistic" methods, which seek to uncover truth without recourse to experiment or quantification. Of course, in reality the distinction is not always so clear: there are some examples of scholarship that do not obviously fall in either, or that can be argued to fall into both. Regardless, I believe the distinction is useful and sufficiently unambiguous for our purposes.

You may ask: Why begin a book on instruction modeling, a very specific technique, with such a general discussion of how to determine truth in education? First, I cannot imagine making *any* claims about education unless we have first established the principles of what constitutes adequate evidence. Second, there is so much confusion—and, frankly, wrongheadedness—on this subject, that many friends and colleagues have encouraged me to address it, arguing that it would have value independent of the remainder of this book. And third, the importance of humanistic scholarship in education *is* the central argument for instruction modeling: after all, if we believed

* I will deliberately avoid using the word "scientific" as a synonym for "quantitative," since it is a term that has become very broad (viz., "social sciences") and therefore does not accurately capture the meaning we seek.

that quantitative tools were enough to know what works in education, what value would there be in carefully studying and recreating the detailed, lesson-by-lesson practices of outstanding teachers? Instruction modeling is important precisely because it provides a way to extract and apply humanistic knowledge of what makes for good instruction, and this knowledge vitally supplements what we can know from experiments and numbers.

What Standardized Test Scores Tell Us

Modern standardized tests appeared in the early twentieth century as a tool for sorting World War I military conscripts into officers and enlisted men.[2] They were meant to be tests of general intellectual ability, not specific content knowledge. The schools adopted the approach, and the College Board Entrance Exam—the predecessor of today's Scholastic Aptitude Test (SAT)—began using tests in the 1920s to sort students into colleges. The tests promised a new certainty and objectivity: students with different grades from different schools could now be compared using a single consistent method—and, moreover, one that bore the imprimatur of science.

Standardized *content*—rather than *ability*—tests appeared shortly thereafter and were promoted by the federal and state governments as a way to assess the academic success of students and schools. In the 1980s, some states began mandating these tests and using them to formally evaluate the performance of schools; the No Child Left Behind Act of 2001 extended this mandate to the entire nation, while leaving to states the decision concerning what particular tests to use.

The rise of testing has become controversial, but—controversy aside—it has yielded comprehensive standardized test data for the vast majority of schools, students, core subjects (mathematics, English, science, and social studies), and grade levels (3–12). This massive collection of data can be used to compare any two schools within a state. If we wish to make comparisons between states, the National Assessment of Education Progress (NAEP) conducts regular studies that test large samples of students from each of the fifty states in mathematics and language arts. And if we wish to go even further, and compare different nations, there are two separate projects—the Trends in International Mathematics and Science Studies (TIMSS) and the Programme for International Student Assessment (PISA)—that test dozens of nations every several years.

In short, we are swimming in test data. One might think this would make it exceedingly simple to identify good instruction: those nations, states, and schools that score best on the tests must be the ones that provide the best instruction. So we have only to identify those top scorers and study what they do.

If only! Sadly, those test scores—despite their precision and consistency—depend not only on instructional quality, but also on many other factors. For instance, imagine a school in a small college town showing better test scores than one in a rural part of the same state. The students in the college town might have parents who are historians, engineers, Latinists, and biochemists—who have the training to help with their children's studies, money to pay for tutors and after-school activities, and access to the best summer enrichment programs. The college town school might be better funded from local tax revenues, allowing for better facilities and a lower student-to-teacher ratio. The college town students might be surrounded by peers who have all been raised with the assumption that college attendance is a given. What part of the difference in test scores is due to differences in instructional quality, and what part due to these other factors? It is impossible to perfectly disentangle them, especially because some important factors—for example, the extent of parents' involvement in their children's education—cannot be precisely measured.

The same problem presents itself when comparing across countries. For example, Japanese students have historically done substantially better on mathematics tests than American students. Is this because of better instruction, or because of other factors? In *Education and Training in Japan*,[3] Thomas Rohlen and Christopher Bjork point out the massive difference in total learning time between the two countries: Japanese students spend more days in school, do more homework, and often attend intensive evening "cram schools." By Rohlen and Bjork's estimate, over a seven-year period of schooling, Japanese students spend an additional 3.2 years on task! Couldn't this just as easily account for the difference in performance?

Time on task is only one difference between countries. Different countries have different levels of wealth, education, parental involvement, views on the importance of academics—and a thousand other such things. Instructional quality is merely one among many different factors that determine how a country's students perform on standardized tests.

These complexities do little to restrain the public's enthusiasm for whichever country happens to top the latest international study. For instance, several years ago Finland put in a good showing, matching the group of Pacific Rim countries (including Singapore and Japan) that have for some time dominated the rankings; the consequent euphoria was such that a leading educational statistician told me that the next time someone asked him about Finland, he would punch them in the nose. Finland is, after all, a nation of 5.5 million inhabitants of remarkable demographic homogeneity. Telling the United States to emulate Finland in education is like suggesting that New York City look for inspiration to the excellent traffic reduction initiatives recently undertaken by Ames, Iowa.

This is not to say that the international rankings have no value: instructional quality surely plays a significant role, and *on average* the higher-ranked countries are likely to have better instructional practices. But because of all the other factors that affect a country's test scores, spotting good instruction is not as simple as looking to the top of the ranking tables. The tables offer a hint, but nothing resembling certainty.

What Laboratory Research Tells Us

> I have formed a method by which, it seems to me, I have the means to increase my knowledge by degrees and to raise it little by little to the highest point which the mediocrity of my mind and the short duration of my life will be able to allow it to attain.
>
> —René Descartes, *Discourse on the Method for Conducting One's Reason Well and for Seeking Truth in the Sciences*

As we saw in the preceding section, the problem with using test scores to identify good instruction is the difficulty in untangling the mess of different factors—not only instructional, but also social, economic, and cultural—that determine these scores. Fortunately, there exists a tool—the scientific laboratory experiment—that is designed exactly for the purpose of isolating the effects of one factor on an outcome. The laboratory experiment is a professional untangler.

Scientific experiments have a long history, and were carried out by the ancient Egyptians, the Greeks, and the Muslim philosophers of the middle ages.[4] By the early twentieth century, scientific experiments had played an essential part in the discovery of the principles of mechanics, electricity and magnetism, the existence of elementary particles, the microbial cause of infectious disease, and the laws of chemistry. The technologies spawned by this new knowledge—engines, explosives, vaccines, steel, the telephone, and many more—transformed life.

Science stood victorious, and scientists the celebrated kings of the new order. The humanities took notice; disciplines that had previously been the domain of humanistic scholarship refashioned themselves after the natural sciences. History was now a "social science," the study of governments had become "political science," and a host of new "-ologies"—sociology, anthropology, psychology—aimed to apply the newly triumphant methods to ancient questions.

Education was no exception. In 1879, Wilhelm Wundt at the University of Leipzig opened the world's first laboratory devoted to psychology.[5] In the first half of the twentieth century, an approach to psychology called *behaviorism*—which studied the behaviors of humans and animals as the products of their experiences—appeared and quickly rose to dominance in American academia. Behaviorism was intensely scientific, and by the middle of the twentieth century educational psychology was done largely through scientific experiments.

How exactly does a laboratory experiment in education look? Typically, we begin by identifying the cause variable and effect variable we would like to study. For instance, we might ask whether background music improves or harms learning; in this case, the presence of music is the (conjectured) cause, and a change in students' learning the effect. Then, we take a collection of people (our experimental subjects) and divide them into two groups; the first one (the "experimental group") is given a lesson with the presence of background music, while the second one (the "control group") is given the same lesson without the background music. Then, we give both groups the same test to see how much of the lesson content they now know; if there is a significant difference in favor of one group, we take this as evidence that background music indeed influences learning.

There are two main things we must do to ensure that our experiment does not lead us to a spurious conclusion. First, we must make sure that in everything except the factor we are studying (in our example, the background music), the two groups are treated identically. The content of the lessons they receive, the room environment, the lighting, the style of the instructor, the comfort of the chairs—all must be indistinguishable, since otherwise we will have no idea whether the effect was truly caused by the background music or by differences in some other aspect of the experience. This is the reason why such experiments are conducted in laboratories—designated spaces where the environment can be tightly controlled. Experiments that satisfy this condition are called *controlled* experiments.

Second, we must ensure that our two groups of subjects—the experimental and control groups—were similar to begin with. If, for example, the members of one of our groups already know more about the topic of the lesson, or have a greater fondness of music, or are especially enamored of written tests, then any difference in the test results could well be the effect of these group differences and *not* of the presence or absence of background music. This problem does not arise as often in the physical sciences: a chemist need not worry that two batches of ammonia will have different views on chemical reactions, nor does a physicist need to be concerned that an electron had slept badly the previous night. But people vary in countless ways, and it is impossible to know whether some difference between groups—perhaps in an attribute we had never thought of, or in one there is no way to measure—will impact our outcome measure and lead to a false result.

How can we solve this problem? The answer is a brilliant idea attributed to the great American philosopher Charles Sanders Peirce (1839–1914) and popularized by Ronald Fisher (1890–1962), one of the originators of statistics.[6] The idea is to *randomize* the selection of the groups: for each participant in our study, we flip a coin to determine whether to place them in the experimental group or the control group. This solves our problem because of a law of probability called the Law of Large Numbers: this law states that as we randomly pick more and more individuals from a large collection, our chosen group will eventually get closer and closer to the full collection in any given attribute, whether observable (like height) or unobservable (like moral fiber). Thus, randomly selecting our experimental and control groups serves to "equate" them—assuming our initial collection of participants is large enough, of course; the larger our groups, the less the statistical likelihood that they will differ in some important attribute that could distort our experiment's results. In practice, a sample of 50–100 participants is usually enough for the Law of Large Numbers to work its statistical magic.

An experiment that satisfies both of these conditions—i.e., is carefully *controlled* and also has *randomly* assigned subjects to the two groups—is appropriately termed a *randomized controlled experiment*. The randomized controlled experiment—our professional untangler—has proven to be an exceptionally powerful method. One example that affects us all is the development of new medicines: randomized controlled experiments (called "clinical trials") allow researchers to successfully determine whether proposed drugs are effective and safe. Only around a tenth of drug candidates pass the test,[7] so we are most fortunate to have a way to separate the wheat from the (quite dangerous) chaff.

Within education, such laboratory experiments have led to remarkable progress over the last several decades. Many educational questions that were previously controversial—and which could not be resolved one way or the other, since there was no objective way to put them to the test—have been decisively settled. For instance, in our example concerning background music, some had argued that such music was distracting and harmful, while others had maintained that it helped to relax and focus students on the task at hand; both groups would appeal to their personal experiences and to common sense, but they each had their *own* distinct common sense and could arrive at no agreement. When at last experiments were conducted, they revealed that background music was generally harmful to learning, especially when the content matter was complex.[8] The dispute was resolved, with no shots fired. We will see many more examples of this in subsequent chapters.

All in all, experiments have given us a better understanding of what factors lead us to remember what we learn, what layouts are most conducive to learning from books or multimedia materials, what motivates students, what the best timing for practice and review is, and many other such important questions. It is not an overstatement to say that the randomized controlled experiment has transformed education from a *speculative* field—i.e., one in which arguments could be made for certain positions, but no truly definitive proof offered—to a *scientific* one. Now, successive generations are not condemned to endlessly replay the same debates; definite progress in our knowledge of what makes for good instruction is possible.

The Limitations of Laboratory Research

Laboratory experiments in education have enabled tremendous progress in understanding what makes for good instruction. But it is too early to pour the champagne: laboratory experiments in education have several serious limitations (see Table 1.1).

The first limitation of laboratory experiments is that we can never be completely certain that our finding is indeed due to the difference in our "cause variable" between the two groups. For example, if our experiment on background music finds that the without-music group outperforms the with-music group, we *interpret* this as meaning that the background music made the difference. But it is also possible that we had done a bad job of controlling the experiment, and in fact there were other, unintended and unnoticed differences between the experiences of the two groups; perhaps the instructor had accidentally delivered one lesson with greater enthusiasm, or the time of day was different and this ended up skewing the results. It is also possible that the effect was a statistical fluke.

TABLE 1.1. Four Limitations of Laboratory Experiments in Education

#	*Limitation*	*Description*
1	Internal validity/ reproducibility	It is hard to ensure that only the independent variable has changed; reproductions of many psychology studies fail to recover the same findings
2	External validity	Results do not always generalize to contexts outside of the laboratory
3	Cannot test entire courses	It is impractical to test whole courses in a laboratory setting
4	Not everything can be measured	Some essential aspects of good instruction cannot be quantified

In technical language, an experiment whose design does not justify the conclusion is said to have poor *internal validity*. Sometimes, internal validity flaws are obvious; other times, they are subtle and disguised. We can never be completely assured of an experiment's internal validity. For this reason, education researchers give more credence to findings that have been reproduced by *many experiments* conducted by *different researchers*. An experiment that consistently gives the same result when thus repeated is said to be *reproducible*.

But isn't this mere hairsplitting? Wouldn't any reasonable experiment be reproducible? To see if this is the case, an ambitious study was conducted by the Open Science Collaboration led by Brian Nosek of the University of Virginia.[9] The study authors identified a hundred different psychology experiments—all published in leading journals—and then carefully attempted to reproduce each one. They only succeeded in 39% of the cases.

This does *not* mean that scientific findings are worthless. For one, of the 61 percent of reproductions that did not recover the original result, only a small fraction obtained the opposite result; the vast majority simply obtained a neutral (no discernible effect) finding. And there are many other sciences—much of biomedical research included—where individual experiments have low reproducibility, but definite scientific progress nevertheless takes place. The main point is that we cannot take the findings of individual experiments to be conclusive: to have a high degree of confidence in a claim, we must see multiple different well-designed experiments—performed by different research groups—that collectively corroborate that claim.

Unfortunately, academia does not reward replication studies: recognition accrues to those who conduct "novel" work. As a consequence, only some claims boast enough experiments to leave no doubt. In general, we must take great care in interpreting scientific findings, weighing the quantity and quality of experiments in deciding how much confidence to have in the results.

The second limitation of laboratory experiments is that *their findings are not guaranteed to hold beyond the narrow context of the original experiment*. For example, our experiment on background music was most likely performed over the course of two hours in a closed, nondescript room in the basement of a psychology department. Would the result be the same if the course lasted not two hours, but a whole semester? Or if the learning environment was a noisy classroom? Or if the walls were full of distracting posters? Or if the music was quieter? Or if the subject matter were different? Or if the students were older or younger?

Because experiments—in order to be so carefully controlled—usually take place in a laboratory, they are *by definition* removed from the realities of the actual environments where learning takes place. Furthermore, owing to the convenience of conscripting one's students as experimental subjects, the majority of education psychology research is done on introductory psychology students at top research universities.

Constantine Lyketsos, a leading dementia researcher at Johns Hopkins University, has lamented that we have cured Alzheimer's disease dozens of times in mice, but have failed to find a cure that generalizes to humans. Likewise, the frenzy of experiments being conducted in the nation's psychology departments gives us hope of curing ignorance in college freshmen; what parts of our experimentally derived knowledge will translate beyond this group, however, is unclear.

An experimental finding that does not generalize beyond its narrow context is said to have poor *external validity*. The way to overcome this is to perform experiments asking the same underlying question in a wide variety of different contexts: different topics, academic subjects, environments, participant ages, and so on. The greater the variety, the more confidence we can have that the result holds true in general. But, as with direct experimental replications, and for the same reasons, this sort of "different-context replication" is rare.

The third limitation of laboratory experiments is that *it is virtually impossible to test an entire course in a laboratory*. It is simply impractical to ask fifty participants to come to a laboratory every day for a year—or twelve.

Offhand, this may not seem like a serious problem: after all, can't we compare two courses piece by piece, comparing corresponding components? Unfortunately, this is impossible because many courses are designed to be used as a whole; different sections of a course are interdependent.

For example, two different Spanish courses would surely teach new vocabulary words in a different order. Imagine we ran an experiment comparing their chapters on the conditional form and found a significant difference. Is one chapter instructionally better than the other? We cannot be sure, because other things could explain the effect. For example, what if the "worse" chapter assumed children knew specialized vocabulary words introduced earlier in that course, and therefore confused our laboratory students? In this case, the chapter would appear worse simply because it was used out of context of the rest of the textbook.

As a second example, imagine one mathematics textbook that teaches multiplication as repeated addition (i.e., 5×3 is the same as $5 + 5 + 5$) and another that merely has students memorize the multiplication tables without drawing out the connection to addition. If we compare these two chapters in an experiment, we might find that students do better on the memorization one, since it does not assume knowledge of addition—and some of our experiment participants might not have learned addition well. We would conclude that the memorization chapter was better—but is that a good conclusion? If the courses were used in their entirety, the conceptual course students would have studied addition before multiplication and we would be quite likely to obtain the opposite result.

There are many other ways in which different parts of a course can be interconnected. A course might focus on developing general learning skills, which would slow down content learning in the short term but may lead to much better learning in the long term; a course could spend time on "detours" to interesting topics off the beaten path, taking time away from immediate concerns but ultimately giving students greater motivation to master the subject; a course might be especially demanding, making it harder-going but ultimately producing more resilient students.

This third limitation of laboratory experiments is more fundamental than the preceding two. The first two—challenges to internal and external validity, to use the technical terms—can in principle be overcome through even more laboratory experiments. The third limitation, however, cannot be overcome by any number of laboratory experiments. Courses are not merely sums of their components, each interchangeable with the corresponding components of other courses; rather, courses are deeply interconnected systems, and as such must be considered as undivided wholes. The laboratory is incapable of testing whole courses.

At first glance, this may seem to be cause for despair. After all, formal education rarely happens through bits and pieces of content: students in schools, colleges, and even online education programs learn through complete courses. If laboratory experiments will not allow us to determine which of two courses is better, then of what use are they?

Actually, of great use. While laboratory experiments cannot evaluate a course in its entirety, they are perfect for the opposite task: isolating a general aspect of education (such as the presence or absence of background music) and studying its impact. Thus, laboratory experiments allow us to find general principles of good instruction. While they cannot directly tell us which of two courses is better, they can tell us that in general, background music should be avoided, conceptual and computational learning are both important, literature classes are best when they include authentic discussion,* and skills should be practiced in many small installments instead of a single long segment.[10] Knowing these principles—and the hundreds of others uncovered through laboratory experiments—*can* help us evaluate courses: those courses that conform with more of these principles are likely to be better.

This is hardly the black-and-white litmus test we would have liked: different people may disagree about the extent to which a course conforms to a principle, or about the relative importance of different principles. So we have not succeeded in removing all subjectivity. But it is still a substantial improvement over how matters stood before the twentieth century, when there were no *generally accepted* principles of good instruction at all.

The fourth limitation of laboratory experiments is that *they can only address those effects that can be measured.* At first, this may seem like a foolish point: if something can't be measured, how can we be sure it is really a thing to begin with? Something that is important for students' learning must surely leave a measurable trace. As many beginning teachers these days are taught, "If there is no evidence of

* A discussion is called "authentic" when the teacher asks genuinely open-ended questions. For example, "Why did Hamlet hesitate to kill Claudius?" is an authentic question: there is no single "correct answer," and therefore the question invites interpretation and discussion. By contrast, "What is the name of Hamlet's mother?" is not an "authentic" question: there is only one correct answer, and the teacher has asked the question only to verify that students have read the book. Of course, there is a place for both types of questions, but research suggests that better literature classes tend to exhibit a larger proportion of authentic questions.

learning, then there is no evidence of teaching"—by which is meant that good teaching is that which improves students' test scores, and a teacher must continue refining their technique until that effect is achieved.

This way of thinking is natural to us, who have all been raised in the age of quantitative science. We are skeptical of claims that are not backed up by numbers. We regard unscientific thought as the province of charlatans.

But are we right in this? In 1963, American sociologist William Bruce Cameron wrote a passage the last sentence of which would become famous, though—unfortunately for Cameron—usually misattributed to Albert Einstein:[11]

> It would be nice if all of the data which sociologists require could be enumerated because then we could run them through IBM machines and draw charts as the economists do. However, not everything that can be counted counts, and not everything that counts can be counted.

French-American intellectual Jacques Barzun was harsher in his criticism. A leading historian and one of the architects of Columbia University's mid-twentieth-century educational reforms (which instituted a common curriculum based on classic books), Barzun was dismayed at the "magic spell of science" and the "assumption that facts exist only for scientists." He argued that much of human experience—including history, art, and literature—could not be decomposed into pieces and measured with numbers:[12]

> But the same human mind that has created science by the analytical method can work in an entirely different way. The mathematician-philosopher Pascal pointed this out 350 years ago. He called the way of analysis the "geometrical bent." It deals with simple things like angles or straight lines or atoms or molecular pressure. They are simple because they are well defined. . . . The other use, direction, or bent, Pascal called the *espirit de finesse*—we might call it "intuitive understanding." It goes about its business just the other way. It does not analyze, does not break things down into parts, but seizes upon the character of the whole altogether. . . . Now, the things that make up culture are understood and remembered and enjoyed by mental finesse; they are for inspection as wholes, not for analysis and measurement.

Is education entirely like geometry, or is it in some ways like art? A simple thought experiment can help us answer this question. Imagine two English courses, each pursuing the same essential goals: to teach grammar, reading comprehension, and writing. The courses are structured identically, with explicit instruction along with readings and discussions of books. The only difference is in the choice of books: one uses excellent literature, while the other uses books that are equally gripping but that are devoid of any cultural significance or artistic merit. At the end of the course, will one group of students differ measurably from the other? It is unlikely: they will have learned as much grammar, and improved their tested reading comprehension and writing abilities by the same amount; and it would be too optimistic to think we could actually measure any impact on their aesthetic sense. I know of no experiment that can demonstrate the instructional superiority of reading Mark Twain over *Us Weekly*.* And yet, would any reasonable person claim that the difference between our two imagined courses is not instructionally important in ways that go beyond subjective preference?

Lest you think that literature is unique in this, ask any mathematician, historian, or physicist whether their discipline has an essential aesthetic element. Even in mathematics—as precise and quantitative a discipline as ever existed—the intangibles matter greatly. There are such things as correct and incorrect mathematics, but there are such things as beautiful and ugly mathematics, too. The great British mathematician G. H. Hardy wrote in *A Mathematician's Apology*, an essay that is the closest thing to a widely accepted manifesto for the whole profession:[13]

> The mathematician's patterns, like the painter's or the poet's must be beautiful; the ideas like the colours or the words, must fit together in a harmonious way. Beauty is the first test: there is no permanent place in the world for ugly mathematics.

* The closest thing is a fascinating study (*Mere Exposure to Bad Art*, Meskin et al., 2013) that found that repeated exposure to the paintings of John Everett Millais—a critically acclaimed artist of the nineteenth-century Pre-Raphaelite school—increased liking of his works, whereas repeated exposure to the paintings of Thomas Kinkade—a popular artist whose works are described by critics as "so awful they must be seen to be believed"—reduced liking. Of course, this study does not "prove" that Kinkade's works are artistically inferior.

Beyond those things that simply cannot be measured, there is a vast world of things that are *impractical* to consistently measure: Is a course inspiring? Do students come to identify themselves with the discipline? What values do students acquire? What beliefs about their own abilities? What motivation to pursue further study? These questions—and a hundred more—matter a great deal, and may well influence the course of students' lives. Some of these things can be measured—for example, through surveys of students' beliefs—but measuring *all of them* in a given experiment is certainly impossible. And even if we could, how would we interpret the mess of numbers? If a curriculum improved students' motivation by 5 percent but reduced their belief in their potential by 3.5 percent, what are we to make of it?

What Field Experiments Tell Us

Laboratory experiments are excellent tools for identifying general principles of good instruction, but—as we saw in the preceding section—are unsuited to evaluating the instructional quality of specific schools, courses, or programs. One way to overcome this is through *field experiments*, which are experiments conducted in real settings. For example, if a school principal is unsure which of two textbooks is better, she could ask a teacher to use one textbook with one group of students and the other textbook with another group, with students randomly assigned to the two groups.

Field experiments in education have two major advantages over laboratory experiments: first, because they occur in a natural setting—rather than the artificial setting of a laboratory—their findings are more likely to generalize to other real situations; and second, a field experiment can be conducted over an entire course lasting a semester, a full year, or even longer.

Of course, not all field experiments are created equal. Some of the most rigorous experiments are *school-level randomized experiments*, in which often dozens of participating schools are randomly split between experimental and control conditions. Other field experiments are less robust, occurring on a smaller scale (e.g., within only a handful of schools or classrooms) or replacing randomization with an easier alternative (most commonly, nonrandom selection of demographically similar students to serve as a control group).

It would seem as though we have arrived at the answer to all of our most pressing educational questions: simply conduct a school-level randomized experiment, and the answer yields itself up. Alas, even this gold standard has limitations.

The first limitation is rather mundane: school-level randomized studies are very expensive. Recruiting dozens of schools, offering a program for free (essential if schools are to agree to be randomly assigned to one or the other group), monitoring the experiment's progress, and gathering data add up to a sum in the millions of dollars.

Second, even a very large study is still just *one* study, and as such might not generalize as far as one might like. Even if the participating schools are fully nationally representative—a rare circumstance—there may be idiosyncrasies in the way the program was implemented. To use the technical term from before, we are not assured of *external validity*. This can be addressed by conducting multiple experiments in different settings, but—owing to the cost—this is quite rare in practice.

The third limitation is more fundamental: field experiments can help us compare programs or approaches, but not instructional traditions in different countries or times. This is because—as we will discuss at greater length in the next chapter—faithfully implementing one country's (or time period's, for that matter) instructional approach in another is exceptionally difficult. At first, this may not seem like a serious limitation: after all, is there not enough to learn from different programs within one country? However, as it turns out, nations often have substantial homogeneity in their instructional methods, whereas across national borders the differences are usually striking. Thus, if we confine ourselves within one nation in our search for good instruction, we will be searching within a very narrow tract.

The fourth limitation of field experiments is a familiar one: just as in laboratory studies, there are some things that matter but cannot be measured. Arguably, this is an even greater problem for field experiments: laboratory experiments usually test specific principles, which can be narrowly defined; by contrast, field experiments are often used to evaluate whole programs, which naturally have effects on many different facets of learning.

The Nefarious Alignment Effect

The fifth limitation of field experiments is the most important: standardized tests—some one of which is invariably used as the "primary measure" in any field experiment—are not the simple, unambiguous measures the general public holds them to be. Psychometricians—experts in educational measurement—are aware of the difficulties of interpreting test scores, but members of the general public, many school teachers and administrators, and even many education researchers are not.

If we wish to measure the length of a table, we use a yardstick. Another yardstick—assuming it is not defective—will report the same length when applied to the same table. Most people think of tests in the same way: instead of measuring the length of a table, we are measuring a student's knowledge of mathematics. As with length, we obtain a number. And if two tests are both well-designed tests, then these two numbers will agree. Unfortunately, this is far from the truth. Harvard psychometrician Daniel Koretz writes in his outstanding book *Measuring Up: What Educational Testing Really Tells Us*:[14]

> Of the many complexities entailed by educational testing, the most fundamental, and the one that is ultimately the root of so many misunderstandings of test scores, is that test scores usually do not provide a direct and complete measure of educational achievement. Rather, they are incomplete measures, proxies for the more comprehensive measures that we would ideally use but that are generally unavailable to us. . . . *Tests are generally very small samples of behavior that we use to make estimates of students' mastery of very large domains of knowledge and skill* [emphasis added].

What does this last sentence mean? Take the example of fifth grade mathematics. We wish to know, "which of two students knows fifth grade mathematics better?" The whole of "fifth grade mathematics"—which psychometricians would call the *construct* in this measurement challenge—is very, very large. There are millions upon millions of different problems we could ask children to solve; these problems differ in their content coverage (multiplication versus addition, fractions versus whole numbers, geometry versus arithmetic . . .), the nature of their difficulty (conceptually hard problems, computationally hard ones, ones that require creative thinking . . .), the specific wording of the problems, and many other dimensions. When we give a test consisting of only fifty questions, there is no way that this test will uniformly cover *everything* included in the whole of fifth grade mathematics: the test will necessarily *sample* a very small piece from the whole. If a student happens to be better in the areas the test samples, then the student's score will be overstated; if worse, then understated. Given this, it is not difficult to imagine two tests that disagree about which of two students knows fifth grade mathematics better.

These are not just theoretical considerations: to those who have dealt with educational data firsthand, they are a daily reality. Koretz gives some startling illustrations. For example, it has long been observed that when states change their high-stakes tests, the next several years usually see rapid annual increases in scores on the new exams. Are students actually improving, or are schools merely getting better at targeting their instruction to the parts of the subject sampled by the new test? To find out, Koretz and colleagues had students in 1990 take a high-stakes test that had been discontinued and replaced by a new test in 1987.[15] They found a remarkable pattern: while schools had improved their performance on the new test by half a grade equivalent* over the four intervening years, their performance as measured by the old test had deteriorated by the same amount!

I will not belabor the point through further examples, as Koretz gives plenty in his fine book. Suffice it to say that half of a grade equivalent is not unusual: even tests designed to agree with one another can easily differ by this much or more because of subtle differences in content emphasis.

This phenomenon has profound implications for field experiments in education: for example, if we conduct an experiment comparing two programs and find a difference of half a grade equivalent or less, we have no idea whether there is a genuine difference in instructional quality, or whether one of the programs is merely closer in its emphases to the test we used. Put in more technical language, there

* "Improved by half a grade equivalent" means that the students did as well on the test as they would be expected to have done if they had received an extra half grade of schooling.

is a "measurement error" (analogous to the margin of error of political polls) that is on the order of half of a grade equivalent.

An error of half a grade equivalent is massive. It is difficult to imagine instruction so effective that it will improve students' knowledge by one and a half grade equivalents for every grade of schooling, and—despite what the occasional inspirational film will have us believe—any sensible person involved in teaching knows that education is a slow process, and good instruction does its work in small increments that add up to big differences only over many years. Research bears this out: even the most successful programs boast only fractional grade-equivalent effects in well-designed yearlong field experiments.[16]

What does this mean for field experiments? It means that in practice, the findings from one test are insufficient to draw definitive conclusions about the difference—or lack of difference—in instructional quality between two programs or approaches, no matter how well-designed the experiment. Such findings *can* tell us that one program will consistently better prepare students for a specific test, but a different test—even one covering the exact same content standards—could well show the opposite result.

Educational measurement using field experiments is nothing like measuring the length of a table. If we seek a physical analogy, a better one would be determining whether a barn is on the whole well-painted, under the constraint that we can judge only by viewing an up-close photograph of one small portion of the barn's exterior. Our photograph will not constitute the sort of evidence we would bet our lives on. But nevertheless, it would still give a hint as to the quality of the whole. And our confidence would be higher yet if we managed to get our hands on two or three different such photographs. And if, for example, we learned that no one had ever taken a photograph of the barn that showed good paint quality, we would be rightly reluctant to retain the painter for our own home improvement projects.

The discipline that seeks to determine the instructional quality of programs is called *educational program evaluation*. Program evaluators have a difficult task: as the preceding discussion shows, there is no simple, decisive way to evaluate a program. Even the strongest experimental design—the large-scale randomized study—yields evidence that is helpful, but far from conclusive. Program evaluation is less like a pregnancy test and more like art criticism.

This can be frustrating to those who seek easy answers. But those who seek easy answers would be well-advised to avoid education in any case.

Understanding this, good program evaluators gather every scrap of evidence they can obtain. They begin by studying the program's design to determine whether it conforms to the principles of good instruction identified by laboratory research; this helps to understand whether the program is a priori capable of having a positive effect. Then, they investigate the process of how the program is used: how much time students actually spend on the program, how teachers engage with it, and what the classroom environment is like. Next, they look at a variety of field studies of the program. Studies with stronger designs—such as large-scale randomized studies—are given more weight, but all sensibly designed studies are considered, as they provide invaluable further evidence. Returning to our barn analogy, a weaker study is like a blurrier photograph: less useful on its own than a crisper one, but nevertheless valuable, as each additional photograph gives us information about a new angle on the barn surface.

And perhaps most importantly, when program evaluators design and conduct their own field experiments—their central work—they rely on *multiple measures*. These can include various standardized tests, student course grades, behavior, attendance, classroom observations, student behavior, teacher attitudes, and a wide range of other such things. The result is not a simple yes or no, but rather an impressionistic picture consisting of many parts, some clear, some murky, some consistent, some contradictory. Sometimes, it is enough to draw a sure conclusion; but usually, the situation is more nuanced. As SRI International's Barbara Means and Jeremy Roschelle—two leading evaluators of education technology programs—explained in an interview with the news outlet EdSurge, the only truly definitive answer to most questions about effectiveness is that "it depends."[17]

The Best Approach: Combining Quantitative and Humanistic Methods

> "We dissect flies," said the philosopher, "we measure lines, we combine numbers, we agree about two or three things which we do

understand, and we disagree about two or three thousand which we don't."

—Voltaire, *Micromegas*

This state of affairs is normal for academics, who are famous for qualifying every claim. But for most, uncertainty is an uncomfortable condition. Koretz recounts a time he was involved in congressional testimony concerning the evaluation of a federal program, and the evaluators were unable to determine whether the impact was positive or negative given "any reasonable estimate of the margin of error."[18] The incensed subcommittee chair asked, "What the hell is this 'margin of error' stuff? Doesn't that mean you don't know what the hell you are talking about?"

At this point, the reader may feel like the congressman: we have introduced a procession of remarkable quantitative tools for identifying good instruction, only to then point out the flaws of each. Now is a good time to take stock of our position. What are our major findings thus far?

1. Because instructional quality is only one factor in test scores, the schools, states, and nations with the best instruction are not necessarily the ones with the highest test scores.
2. Randomized laboratory experiments are excellent untanglers, allowing us to understand how a single variable impacts instructional quality. However, laboratory experiments cannot evaluate entire courses; they are best suited for identifying *general principles* of good instruction.
3. Field studies (with large-scale randomized experiments chief among them) can be used to evaluate whole courses or instructional programs. However, there is no single experiment or measure that gives a definitive answer: our conclusion must be pieced together from an impressionistic mix of studies, tests, and other indicators, and is rarely definitive.
4. There are certain important facets of education that are not in practice measurable, either in the laboratory or in the field. For practical purposes, they fall outside of the domain of quantitative scientific inquiry.

In other words, the importing of quantitative methods into education scholarship has been a great improvement for the better. We now have reliable, largely objective tools for establishing principles of good instruction. We also have methods for evaluating specific courses or programs; while the picture they paint is an impressionistic one, it is still a great improvement over how things stood before: a lack of any objective methods, leaving each observer to paint whatever picture they wished in their imagination.

At the same time, our new quantitative tools still leave major gaps, and it seems unlikely that these gaps are simply waiting for the next big leap in research methods. Instead, we should accept education for what it is: a field that straddles the divide between the sciences and the humanities. Just as it is ludicrous to suppose that someday literature professors will support all of their claims with experimental evidence, so too there is a part of educational inquiry that will forever be beyond the reaches of quantification.

But, just as in disciplines such as history, law, or art criticism, this does *not* mean that all is subjective and each opinion equally valid. There is bound to be more rancor and less agreement, a greater variety of differing views, and a perpetual return to age-old questions; but there will nevertheless be many issues in which a consensus among the well-informed emerges. Historians may not agree on as many things as scientists, but few would argue against the value of their knowledge in guiding political and social decisions.

Owing to the general dominance of science in academia, humanistic work is increasingly uncommon in the nation's leading colleges of education—a disturbing trend, given the importance of the questions only it can address. When it does take place, it must often wear the outer trappings of empiricism—such as scientific-sounding terminology—to avoid detection.

When it does take place, what forms does humanistic scholarship in education take? Table 1.2 offers a classification. Arguably the oldest form is the *philosophy* of education; one of the questions this field is especially able to take up is the matter of *what* should be taught in the first place: should students be prepared for specific trades, or given a broad education that will enable them to make the decision on their own as adults? Should the great works of the past be taught, or newer works that may have more relevance to today's particular preoccupations? Is it more important to transmit concrete knowledge, or to attempt to develop students' general abilities

TABLE 1.2. Forms of Humanistic Work in Education

#	*Form*	*Example questions it can address*
1	Philosophy	What should be taught? Should different students be given educations with fundamentally differing goals?
2	History	What educational reforms were attempted in the past, and what was the outcome?
3	Ethnography	How do education systems work in other times, countries, or contexts?
4	Criticism	Do certain curricula accurately reflect the ways of thinking that are recognized as important by leading practitioners of a discipline?

and characters? There are certainly some quantitatively verifiable facts that are relevant to these questions, but they are just as much questions of *values*.

A second form is education *history*. History tells us of the particular educational practices of past societies, the intended and unintended consequences of educational reforms, and the evolution of views on teaching and learning. It is difficult to overstate the value of this knowledge for those who labor in education. One example is the short but excellent history *Teachers and Machines: The Classroom Use of Technology Since 1920* by Stanford's Larry Cuban.[19] This work chronicles the successive waves of much-hyped education technologies of the last century: instructional film in the 1920s, radio in the 1930s, television in the 1950s, and computers in the 1980s. Each technology was rung in with great public fanfare, designed with little consideration for the hectic realities of teachers, used little, and ushered out quietly. The history has many invaluable lessons for would-be designers and users of today's instructional technologies—for example, that the first and simplest step in assessing the value of any technology program should be to look at usage patterns to see if it is actually being accepted by teachers and students. And yet, you could follow Diogenes in searching with a lantern in broad daylight for a blended learning practitioner who is properly familiar with this history of his own trade.

A third form is *ethnography*. This branch of anthropology studies the practices and beliefs of social and cultural groups. In education, this typically means studying teachers and students in specific classrooms, schools, or nations, seeking to understand in detail how education takes place in the community being considered. One exemplary and influential work in this genre is Liping Ma's *Knowing and Teaching Elementary Mathematics*.[20] This book was the product of extensive interviews Ma conducted with several dozen American and Chinese teachers of mathematics; she chronicled and compared their mathematical knowledge, approaches to thinking about mathematical ideas, instructional practices, beliefs about the role of mathematics education, and manner of preparation for teaching. Lucid, detailed, and direct, Ma's work is an excellent illustration of the value of ethnographic work in education. Quantitative experimental studies are simply not the right tool for giving the sort of broad, thorough understanding of the differences between instructional traditions that a study like Ma's provides.

The fourth form we will mention here can be termed *criticism*, by analogy with art or literary criticism. Criticism is the humanistic alter ego of program evaluation; as Jacques Barzun argues in his essay "What Critics Are Good For," the most fundamental purpose of art criticism is to reveal the best artistic products so that they can be discovered, understood, and enjoyed: "Critics must be on hand to give access to art."[21] An educational critic might do the same, reviewing courses, textbooks, schools, and even entire national education systems.

While an educational critic lacks the quantitative tools of the program evaluator, she is able to address the intangible aspects of an educational program, such as its aesthetic qualities, choice of emphases, and faithfulness to the spirit of the discipline. Given the importance of these—and other similarly intangible—facets of education, the paucity of educational criticism is unfortunate. Unlike the other three forms of humanistic scholarship we mentioned—philosophy, history, and ethnography—educational criticism is not an established, broadly recognized field of work. I do not know what accounts for this: we enlist film critics to spare us the discomfort of two hours at a bad movie, but neglect to do the same where the years-long education of our children is concerned.

One group particularly well-suited to performing educational criticism are academics from the subjects in question: mathematicians for mathematics education, historians for history education, scientists for science education. University professors are not always the best judges of what will be successful with school-age students, but as a group they undoubtedly have the best understanding of the essence of their respective subjects. They are, after all, the custodians of their disciplines. This perspective is an invaluable one, since excellent instruction is not only accessible, but also reflective of the spirit of the underlying field. I have seen many examples of this in mathematics: while mathematicians often misjudge the preparedness of students and teachers for this or the other concept or material, they generally have a much better ability than any other group to determine whether a certain course, textbook, or method is mathematically sound.

Sadly, university professors are rarely involved in matters of school instruction, except as occasional participants in state- or national-level debates on what content should be taught.* This is caused partly by their own disengagement, and partly by the lack of awareness on school officials' part of the value professors could provide. And surely a role is played by professors' reluctance to make enemies by criticizing specific textbooks or instructional approaches.

As a result, a stunning divide has formed between mainstream school curricula and university professors' understanding of the meaning of their disciplines. Perhaps the best-known illustration of this is the story of when Richard Feynman, a Nobel laureate and one of the great physicists of the twentieth century, was invited to review school textbooks as a member of California's State Curriculum Commission. Feynman was scandalized as much by the quality of the books as by the outlandish process used for selecting among them. He wrote about it in his famous memoir, *Surely You're Joking, Mr. Feynman:*[22]

* The exception would be professors of education, who provide training to future teachers, write school curricula, and sometimes work with local school districts. However, the point here is that there is value in the involvement of professors from the disciplines (such as mathematics, the sciences, and literature), as they provide an important additional perspective.

> It was a pretty big job, and I worked all the time at it down in the basement. My wife says that during this period it was like living over a volcano. It would be quiet for a while, but then all of a sudden, "BLLLLLOOOOOOWWWWW!!!!"—there would be a big explosion of the "volcano" below.
>
> The reason was that the books were so lousy. They were false. They were hurried. They would *try* to be rigorous, but they would use examples (like automobiles in the street for "sets") which were *almost* OK, but in which there were always some subtleties. The definitions weren't accurate. Everything was a little bit ambiguous—they weren't *smart* enough to understand what was meant by "rigor." They were faking it. They were teaching something they didn't understand, and which was, in fact, *useless*, at that time, for the child.

This all illustrates the special contribution that humanistic approaches are able to make toward identifying good instruction. The philosophy of education helps to understand what should be taught in the first place, the history of education helps to foresee the full consequences of instructional reforms, ethnography sheds light on instruction by providing detailed descriptions of the educational systems of schools or nations, and criticism evaluates the less tangible aspects of courses and instructional methods. Collectively, these approaches go a long way toward filling in the gaps in what quantitative methods can tell us about good instruction.

I am reminded of an old Russian joke. The three musketeers are out riding when they run into the three bogatyrs, mythical Russian knights of yore often portrayed as rough, barrel-chested giants. The groups quarrel, and Atos challenges one of the bogatyrs to a duel. Atos asks for a piece of chalk, and proceeds to place a small X over the bogatyr's heart. "What's that for?" asks the bogatyr. "This is where I will pierce you with my foil," replies Atos. "Oh, I get it," says the Bogatyr. "Dobrynya, sprinkle this one with chalk and bring me my mace." Quantitative methods are like the foil, while the humanities are like the mace: what they lack in precision they make up in coverage.

Put another way, quantitative methods give us a high degree of confidence about a small number of matters, while the humanities give us a lower degree of confidence about a much greater number of matters. Numbers give *more certain* knowledge, while the humanities give *richer* knowledge. This is why the best approach is to consider *both* quantitative and humanistic evidence, coming to a conclusion that benefits from the particular strengths of each.

Of course, this is not to say that all humanistic evidence is equally valuable. To assert this would be tantamount to saying that all opinions are valid, and each observer—no matter how ignorant—is entitled to his views. Just as quantitative evidence differs in quality, so too humanistic scholarship can be good or bad. A historian who does not conduct thorough research and does not base her conclusions on careful reasoning is no different from a quantitative researcher who designs flawed experiments and bungles his statistical analyses. In both cases, we must use our judgment to determine whether the evidence is admissible: Does the work appear sound? Are the arguments grounded in logic and supported by fact? Has the work been peer reviewed or otherwise evaluated by experts? Is the author trained in a relevant discipline? These questions are equally relevant to evaluating quantitative and humanistic work. In both cases, informed people can tell apart good from bad.

Figure 1.1 provides an overview of the process we should follow to identify good instruction. All told, if we wish to know whether a course, textbook, or program represents good instruction, we should look at whether it conforms with scientifically supported design principles; review all available field studies of its effectiveness; consider multiple measures, including as many different standardized tests and other instruments as possible; inquire after the opinion of experts—including professors specializing in the subject itself; compare the program with the qualities of good instruction as revealed by ethnographic studies; and evaluate the program's approach in light of the history of similar programs of the past. And then, we must weigh all of these things together, using our good judgment to reach a conclusion. This may be less satisfying than a simple yes or no, but some fields are inherently murkier than others. We cannot hope to have the same certainty in our conclusions in art history as in chemistry, and education sits somewhere in the middle. But a degree of murkiness should not prevent us from using the best methods possible to get closer to the truth. And the evidence is clear: our best chance of knowing good instruction when we see it is a combined quantitative and humanistic approach.

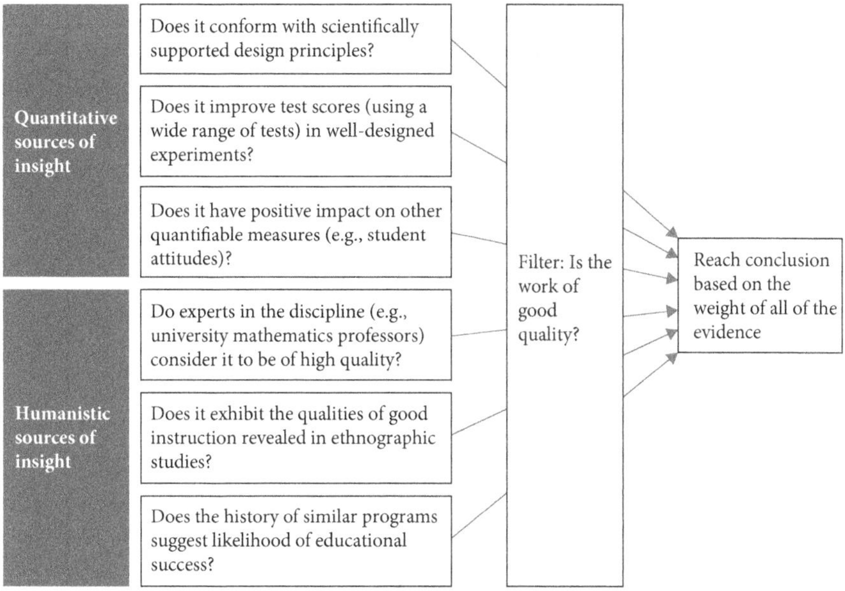

FIGURE 1.1 The process to identify good instruction.

Despite the compelling reasons, it is rare to find discussions of good instruction that do this. Perhaps the closest thing is "mixed methods" research, which combines a variety of methods and measures to form a holistic picture of an educational program or question. However, the "mixed" in "mixed methods" refers to a mingling of *quantitative* and *qualitative* methods, with both still hailing from the experimental, not the humanistic, tradition.

Good Instruction Comes in Systems, Not Isolated Practices

One of the most influential works of ethnography in education is the series of TIMSS video studies, popularized through the bestselling book *The Teaching Gap* by James Stigler and James Hiebert.[23] Conducted within the umbrella of the Third International Mathematics and Science Study (TIMSS, since renamed the Trends in International Mathematics and Science Studies), the video studies gathered recordings of eighth grade mathematics lessons in different countries and then analyzed the lessons in great detail. The first study, done in 1995, used 231 complete lesson recordings from the United States, Germany, and Japan.

Following the ethnographic tradition, the researchers sought to understand everything that distinguished the instructional approaches of the three nations, including teachers' practices, beliefs, values, and professional preparation. Beyond observing and discussing the videos, they developed what are called "coding" schemes—procedures that different individuals can follow to numerically score videos. For example, lessons could be coded for the degree of mathematical difficulty of the material covered. To ensure that scores are not subjective, researchers verify that different coders agree with a frequency at least 80% better than chance. Coding is a way to inject a dose of empiricism into ethnographic (or other essentially qualitative) work; it can be used to confirm the insights gleaned from less structured analyses.

Based on the findings of the video studies, Stigler and Hiebert argued an important claim: that teaching was *systemic*. By this they meant that teaching was not merely the sum of disconnected practices:[24]

> Teaching is a *system*. It is not a loose mixture of individual features thrown together by the teacher. It works more like a machine, with the parts operating together and reinforcing one another.

They give several examples in *The Teaching Gap*, but I will offer an example from my own experience: the role of homework in different countries' instructional traditions. American mathematics teachers most frequently use homework for *extra practice*: the typical homework assignment consists of a collection of similar problems, usually procedural, over the material of the day. By contrast, Russian mathematics teachers view homework as a method for *review* (i.e., for refreshing students' knowledge): their typical assignment offers a smaller number but greater variety of problems, touching on the day's material but also incorporating relevant content from past parts of the course. The nature of the homework assignments influences the rest of the course; for example, Russian teachers can plan the next day's lesson with the assumption that the material reviewed in homework has been refreshed in students' minds.

Given these radically different purposes of homework in the two countries, it is meaningless to ask: "Is homework, in general, a good practice?" One can only ask whether, in the broader context of American mathematics courses, the additional practice afforded by homework is a good thing; and whether, in the broader context of Russian mathematics courses, the review afforded by homework is a good thing. In other words, homework contributes to the overall quality of instruction only *as a component of an interconnected system*. As Stigler and Hiebert put it:[25]

> If teaching is a system, then each feature, by itself, doesn't say much about the kind of teaching that is going on. What is important is how the features fit together to form a whole. How does one feature connect with the next one? How does an activity near the end of the lesson link back with one at the beginning? This is a very different way to think about teaching. It means that individual features make sense only in terms of how they relate with others that surround them. It means that most individual features, by themselves, are not good or bad. Their value depends on how they connect with others and fit into the lesson.

Thus, good instruction is not merely that which satisfies some checklist of desirable properties; even if there are some universally desirable properties, they tell at best only part of the story. To really know whether instruction is good, we must study it in full detail as a system, understanding each component and how it relates to the rest.

An apt analogy is an engine. To know whether an engine is good, it is insufficient to evaluate the separate parts: we must know how they work together. An engine may have excellent pistons, brilliant valves, outstanding spark plugs, and remarkable cylinders—but if they are assembled badly, it would nevertheless be a bad engine. Moreover, the quality of the individual components can really only be evaluated in context of the rest: what serves as the best possible piston in one engine would be ruinous in another.

In the preceding section, we saw many reasons why humanistic approaches are essential for identifying and understanding good instruction. But this observation—that teaching is systemic—offers the most compelling argument of all. Ethnography—and, to an extent, educational criticism—is uniquely well-suited to understanding whole instructional systems. Quantitative scientific methods allow us to isolate and study individual features, but ethnographic work allows us to study entire instructional systems—with all of their complexities and interconnections—as they occur in real life. Ethnography is crucial in identifying good instruction, and ethnography is again crucial in studying it to understand what makes it successful.

Common Misconceptions

> As Lightning to the Children eased
> With explanation kind
> The Truth must dazzle gradually
> Or every man be blind—
>
> —Emily Dickinson

The relative strengths and weaknesses of quantitative and humanistic methods are not widely understood. This is not surprising: as the preceding discussion shows, there are many subtleties and few straightforward prescriptions. What *is* perhaps surprising is the very wide range of different misconceptions one encounters.

The most common misconception is the simple conflation of test scores with instructional quality. Many politicians, philanthropists, businesspeople, and even some school administrators assume that the best teachers, schools, and national education systems are those which post the highest test scores.

Second, on the opposite end of the spectrum, there are those who discount the value of education research altogether, holding it to be nonsense. Those who hold this view often think that the answers to educational problems are obvious, and research on what makes for good instruction is neither possible nor needed. Given the implicit anti-intellectualism of this position, it is ironic that many of its holders are scholars from other fields, which have historically often held education departments in disdain. Even deep thinkers like Jacques Barzun were not immune to this view: Barzun, when he wrote of it, enclosed the phrase "educational research" in quotation marks.

The various examples I have given of quantitative education research suffice to reveal the folly of such a stance. We may think that good education is common sense, but one person's common sense rarely agrees with another's. If a certain view can be experimentally tested, it would be a great missed opportunity to refrain from doing so. Laboratory experiments may not answer all of our questions, but they do help to tilt the scales when there are disagreements on the principles of good instructional design. Field experiments may not give conclusive answers on the quality of educational programs, but they do provide snapshots that can help us infer the true state of things. Why forego this knowledge?

A third common misconception is that quantitative methods can be sufficient and conclusive in identifying the principles and instances of good instruction. This misconception comes from an exaggerated sense of the precision of standardized testing, and also from a lack of understanding of the extent and importance of the intangibles that cannot be measured at all. Even education researchers often hold this misconception, particularly ones who conduct laboratory experiments but have not seen firsthand the inherent murkiness of field studies. Once, I heard a layperson ask an education researcher what it would take to know if a program is effective; the researcher replied that it would take one well-designed large-scale randomized study. While the spirit of empiricism is laudable, in education a simplistic view of the strength of the lone experiment will lead us far astray.

Fourth and last, there are many who write off humanistic approaches to education. This position is rarely held explicitly, but rather is more of an underlying, unquestioned belief that only numbers can reveal truth. Holders of this view regard educational histories not as guides to action, but as curios—or, at most, sources of illustrative stories; they may enjoy ethnographies of education, but would not deliberately seek them out for insights on what makes for good instruction; and they give little weight to educational criticism, considering it to be nothing more than subjective opinion.

What of it? Does it matter that there are so many misconceptions? Is it important, in the grand scheme of things, for more people to understand the subtleties of how to identify good instruction? The answer is a resounding *yes*: instructional decisions are made by teachers when they plan and deliver lessons, by district administrators when they select textbooks and courses, by parents when they choose schools for their children, by politicians when they enact local and national policies, and by citizens when they vote. And if we use erroneous methods for making these decisions, the result will be worse education.

These issues also matter for those of us who design online instructional programs. And as we will see in the third chapter, the value of humanistic methods—and ethnography in particular—is one of the strongest arguments in support of the practice of instruction modeling.

Chapter 2

Good Instruction in Mathematics

IN THIS CHAPTER, WE WILL APPLY THE PRECEDING chapter's ideas to an example: the case of mathematics. This will illustrate how different methods of identifying and understanding good instruction work together in practice. Furthermore, it will give helpful background for our upcoming discussion of instruction modeling, since mathematics—and Reasoning Mind's programs in particular—will serve as our main example.

The question of what makes for the best mathematics instruction has been steeped in controversy. In the 1960s, research mathematicians dismayed with the prevailing school mathematics curricula led a national reform now known as "new math," with the resulting textbooks ultimately rejected by students, teachers, and parents.[1] The 1980s and 1990s saw a series of public disagreements between different camps that were so deep and divisive, they became known as the "math wars." In 2008, a presidential committee convened by George W. Bush—called the National Mathematics Advisory Panel, or the National Math Panel for short—attempted to resolve the controversies once and for all, only to end up with a set of recommendations that were more of a compromise than a

Instruction Modeling. George A. Khachatryan, Oxford University Press (2020)

DOI: 10.1093/oso/9780190910709.001.0001

conclusion. And today, the Common Core State Standards in mathematics are attempting to reform instruction, but are meeting with political resistance from advocates of local control over educational issues, conceptual resistance from those who disagree with the standards' vision of mathematics education, and practical resistance from publishers who have little financial motivation to substantially change their textbook offerings.

This controversy stands in contrast to the situation with reading instruction. Until the 1930s, the prevailing approach to reading was *phonics*: students were first taught to pronounce the sounds of individual letters, then syllables, and finally to sound out entire words piece by piece.[2] Researchers on reading discovered that expert readers—i.e., literate adults—used an altogether different method, visually taking in whole words and sentences as they read. They concluded that students should be taught to read in this fashion from the outset, and *whole word* instruction was born and came to dominate school reading instruction. But many doubted the effectiveness of the new methods, and in 1961, Harvard researcher Jeanne Chall systematically reviewed the research evidence and found there an irrefutable case for the superiority of instruction relying primarily on phonics: while phonics was not how literate adults read, it was the right stepping stone to get there. While it took many decades, Chall's conclusions eventually became the consensus. The matter was conclusively settled in 2008 by George W. Bush's National Reading Panel—after which the National Math Panel was subsequently modeled—which presented clear, agreed-on conclusions about the qualities of good reading instruction.

Why the great difference between reading and mathematics? I am inclined to think that the cause lies in the subjects themselves. While we intuitively think of reading as a less precise domain than mathematics, the reality is the opposite: English as a whole is a broad subject, encompassing grammar, composition, literature, and much else; but reading is a far more narrow skill, defined as the ability to make sense of written passages. Being such a concrete skill, reading ability can be measured and conclusively understood through experiment. Mathematics, however, is too broad to be so readily picked apart. This illustrates that even within education, different areas yield to quantitative analysis to differing degrees. Broad subjects like mathematics cannot be understood without heavy recourse to humanistic approaches—and this means that a certain measure of subjectivity and ideological strife is inevitable.

That said, what *can* we piece together about the qualities of excellent mathematics instruction?

Guidance

> He might mention to the young gentlemen that education is a drawing out, not a putting in. The late Lord Brancaster was much addicted to presenting prizes at schools, and he invariably employed this dictum.
>
> —Jeeves in P. G. Wodehouse's Right Ho, Jeeves, copyright © 1934 by P. G. Wodehouse. Reprinted by permission of the Estate of P.G. Wodehouse

As an example, we will begin by considering a broad question that has been at the heart of great controversy, evaluating it both from a scientific and from a humanistic perspective: *Should students be firmly guided as they learn mathematics, or should they be given room to explore the subject independently?*

The same question can be asked for almost any topic: foreign language, drawing, skiing. But it has been particularly prominent in discussions of mathematics instruction. On one side, proponents of *inquiry-based learning* (the extreme form of which is called *discovery learning*) argue that students learn material more deeply when they arrive at mathematical insights on their own; this leads to firmer knowledge, gives students an opportunity to find their individual preferred ways of thinking about mathematics, and conveys the thrill of mathematical discovery. On the other side, proponents of *direct instruction* believe that students should be explicitly taught the established ways of thinking about and doing mathematics; they argue that this is a more efficient and reliable way of transferring knowledge. Of course, this is something of a simplification: in reality, there are many positions in the middle of the spectrum, and few people would occupy the very extreme on either end. Nevertheless, the broad division holds.

Advocates of the two viewpoints have fought bitterly. The "math wars" of the 1990s provide a recent example, as this question was one of the central ones dividing the two camps.[3] Supporters of inquiry-based learning criticized direct instruction as an inflexible approach that underestimated children's abilities, neglected the importance of conceptual understanding, and turned learning into a tiresome chore; meanwhile, opponents of inquiry-based learning argued that, given the thousands of years it took humanity to create mathematics, it was unreasonable to expect each child to accomplish the same in the span of a decade.

Laboratory research has shed considerable light on this controversy—and in this light, discovery learning does not come off well. Studies across many decades and many domains (including mathematical problem-solving, programming, and others) found that guided methods—in which teachers give explicit instruction, direct students' work, and give more frequent feedback—consistently lead to faster and deeper learning than pure discovery. In a colorfully titled review article ("Should There Be a Three Strikes Rule Against Pure Discovery Learning?"), eminent psychologist Richard Mayer wrote:[4]

> There is sufficient research evidence to make any reasonable person skeptical about the benefits of discovery learning . . . [but] like some zombie that keeps returning from its grave, pure discovery continues to have its advocates.

Cognitive psychology research has generally found that successful instruction is that which presents the right facts and ideas, and then creates an environment that encourages students to reflect on and absorb them. Discovery learning succeeds in the second task—encouraging students to think—but comes up short in the first task; students participating in discovery learning waste time on dead-end lines of thinking, arrive at incorrect conclusions, and generally "fail to come into contact with the to-be-learned material," as Mayer puts it.

On the other hand, research also strongly supports the importance of teaching methods that promote mental engagement with the material. Direct instruction that merely presents the material—such as a lecture on its own—is generally unsuccessful in getting students to think about the content and form knowledge. Perhaps the best-known work illustrating this principle is research on "self-explanation" by the University of Arizona's celebrated cognitive psychologist Michelene Chi. Chi found that when students were prompted to explain what they were reading out loud—even receiving no feedback whatsoever on the quality or accuracy of their explanations—they understood the material considerably better.[5] In other words, students frequently do *not* spontaneously engage in active reasoning about material they read or hear, but the right instructional methods can elicit such reasoning. Inquiry-based elements—such as having students explain concepts to one another or pose questions about the material—can be one way to accomplish this.

Thus, research comes down on the side of guided forms of instruction, but with the caveat that instruction should nevertheless require students to be mentally active; a certain measure of inquiry-based instruction can be helpful to accomplish this.[6] The great number of experiments that confirm this view—and there have easily been dozens, if not hundreds—paint a consistent picture. This is an excellent example of how laboratory research can give clarity on a question that otherwise could be debated ad infinitum.

What does a humanistic approach have to say on this question? Ethnographic comparisons between different countries' education systems, such as *The Teaching Gap*, reveal that there is more than one pattern for excellent mathematics instruction.[7] Japan, generally held to have outstanding school mathematics, makes heavy use of inquiry-based learning: teachers structure lessons around problems, solved by students in groups and then discussed by the whole class. Students are led to make their own conclusions, although teachers guide the discussion along a carefully planned path. This is not pure discovery learning, but it is reasonably far into the inquiry-learning side of the spectrum. But other high-achieving nations differ; for example, Stigler and Hiebert note that "teachers in the Czech Republic and Hong Kong spend much of their time teaching the whole class, whether through lecture or recitation."[8] All of the high-achieving nations, however, "accomplished the *engagement of students in active struggle with core mathematics concepts and procedures*."[9] This corroborates the experimental evidence: there must be enough teacher guidance (in some form) to ensure that students are focused on the right things, and there also must be some ingredient—whether through elements of inquiry-based learning or something else—to ensure students make active sense of the material.

But there is also some evidence from the humanistic perspective that calls the rejection of discovery learning into question—most notably, the Moore method of instruction. Named after its originator, University of Texas mathematics professor Robert L. Moore (1882–1974), the Moore method is as pure an expression of discovery learning as you can find.[10] In Moore method classes, students are given worksheets listing the mathematical results (theorems and the like) of the course; they must prove each theorem independently, developing the necessary ideas themselves as they go. The class discusses each item to come to an agreement, while the instructor passively observes. Some Moore method instructors intervene as needed, but in general this is avoided. Moore himself famously said, "That student is taught the best who is told the least."[11]

The Moore method has been used with great success in graduate mathematics programs, undergraduate courses, and even high school programs for mathematically advanced students. For example, Russia's most famous mathematics school—Moscow School #57—has for many years used it to teach calculus to its most advanced students. The many alumni of the method have extolled its virtues. By all accounts, Moore method classes give students the thrilling sensation of really *doing* mathematics, a deep and lasting understanding of the material covered, and a mathematical self-reliance that serves as excellent preparation for higher-level mathematics courses. Having taken Moore method classes myself, I have experienced this firsthand.

Does this mean that quantitative and humanistic approaches suggest opposite answers to our question? In fact, no. Those situations where the Moore method has been consistently successful seem to be quite special: small classes, outstandingly qualified instructors, and students who are mathematically advanced and exceptionally motivated. These are not the circumstances under which discovery learning was studied in the laboratory; in fact, in *those* circumstances—specifically, with students of more ordinary motivation and preparation—the humanistic evidence is in full agreement with the quantitative. Furthermore, the main effects of Moore method instruction—as claimed by its advocates—are certainly not more efficient learning of the specific content, but rather lasting changes in what mathematicians call "mathematical maturity": an amorphous notion that might be impossible to measure, especially in a laboratory.

Weighing the totality of this evidence, quantitative and humanistic alike, we are led to conclude that pure discovery learning is a poor choice for general mathematics education; that techniques which require students to actively think about the material are nevertheless essential, and may come in the form of (guided) inquiry-based learning; and that in certain special situations—especially where highly motivated students are concerned—there is likely substantial value in discovery learning (specifically, as Moore method instruction) as one component of a program of study.

This discussion illustrates the beautiful interplay between quantitative and humanistic approaches. Far from standing in contradiction, these approaches complete one another, helping to form a clearer and fuller picture of the question being considered. The evidence furnished by each source prevents us from drawing incorrect generalizations from the other source. For example, a person who relied on the experimental results but ignored the humanistic evidence might oppose Moore method courses altogether, thus missing their potential benefit in forming future mathematicians; on the other hand, a person unfamiliar with the research evidence might overgeneralize the benefits of the Moore method, allowing their enthusiasm to carry them into forms of instruction ill-suited to widespread classroom use.

It would be inaccurate to say that the question we just discussed is *the* central question in mathematics education. It is certainly *one* of the most controversial and important questions, but there are many of similar standing: Which is more important, concepts or skills? Should students be taught pure mathematics (as understood by mathematicians) or practical mathematics for their daily lives? How does one teach "problem-solving" skills? Should proofs be taught? And many more. Our detailed discussion of one question helped to illustrate how we can go about combining quantitative and humanistic approaches, but doing the same with all of the most significant questions would take a book—if not a bookcase. But we would nevertheless like a sketch of what the evidence as a whole says about good mathematics instruction. In the following several sections, we will discuss several qualities that are, in my view, particularly significant in characterizing good mathematics instruction.

Genuine Understanding

Vergil famously wrote in the *Georgics*, "Happy is one who is able to understand the causes of things."[12] The line could be taken as a motto for mathematics. In good mathematics education, students do not merely memorize, but rather achieve—whenever possible—real understanding.

For example, it is widely known that in order to multiply two fractions, we must multiply the numerators and denominators separately. But *why* is this the case? Why does *that*—and not some other procedure—give the correct value? If we cannot answer this question, then we may be able to *perform* the multiplication of fractions, but we cannot say we *understand* it.* It is like reading a line of poetry without grasping the meaning. Mathematics education that does not give the place of honor to understanding is like literature education in which children learn to read books aloud with no attempt at comprehension.

The value of understanding is well-supported by experimental evidence. Things that are understood are better remembered.[13] Moreover, knowledge is far more flexible when it comes with understanding: it can be applied to new situations, not just familiar ones.[14] For example, a student who knows superficially how to multiply fractions might be stumped when asked to divide them; but if a student truly understands why fractions are multiplied as they are, she will be far better equipped—should the circumstances require it—to work out the procedure for division of fractions on her own. Psychologists refer to this concept as *transfer of learning*. Knowledge with understanding *transfers further* than superficial knowledge, and experts in any field have knowledge structures that are *deep*—i.e., suffused with genuine understanding—and thus flexible enough to transfer to a wide range of questions.[15]

* Note that understanding and discovery learning are not one and the same: it is possible to understand something without having discovered it by oneself. This distinction is sometimes lost in debates on the appropriate amount of guidance in instruction.

The humanistic case for understanding is stronger still. Every educational critic worth his salt—and also every mathematician—will confirm its importance, and in nations with strong mathematics instruction it is a sine qua non. Understanding is the main rationale for what is known as *curriculum coherence*: a coherent curriculum is one where conceptually prerequisite topics come first.[16] For example, it is impossible to learn decimals with understanding without already knowing fractions (since decimals are just fractions whose denominator is a power of 10); a coherent curriculum would thus teach them in that order.* Nations with good mathematics instruction have coherent curricula, whereas ones with poor instruction often do not.

None of this implies that memorization has no place; many things (the multiplication tables being a classic example) need to be learned by rote, and mathematicians are no enemies of memorization. Leonhard Euler—one of the great mathematicians—had committed to memory the entirety of Vergil's masterwork, the *Aeneid*. Perhaps he had done it out of respect for the poet's admirable views on the value of understanding, but more likely it was because he liked the poetry.

Understanding is not *always* possible: good mathematics education should aspire to impart understanding of every concept on every student, but expecting to always be successful is unreasonable, and sometimes students can get by for the time being with only superficial knowledge of some topic—which is still better than no knowledge at all. But in good mathematics instruction, understanding is always the first and highest goal.

Skills

A major debate in mathematics education has concerned the relative importance of concepts and skills. Some have argued that conceptual knowledge is most important, and excessive focus on practicing skills only serves to dampen students' interest in mathematics; others have argued that skills are the ultimate aim of mathematics instruction,

* By contrast, I have seen examples of (incoherent) curricula that teach decimals first, teaching students to add, subtract, multiply, and divide them by analogy with whole numbers. Thus, children learn the algorithms, but do not understand what they mean.

and students should learn them solidly in order to be mathematically proficient in everyday life.

Here, as in the matter of understanding, experimental and humanistic evidence is in agreement: concepts and skills are both essential.[17] Moreover, they are interdependent; if one is lacking, the other suffers. Specifically, students who do not understand concepts will not be able to remember the skills they learn. On the other hand, students who do not have fluent skills will be bogged down, wasting attention on basic calculations instead of focusing on the new concepts they must learn. For example, a student trying to learn column multiplication will need to do much addition in the process, since addition is an intermediate step in the multiplication of multi-digit numbers. If addition is second-nature to this student, it will not distract him from the new topic. But if the student struggles with addition, then he will be so preoccupied with it that he will not have mental room left to think about and make sense of the new concept of multiplication. For this reason, University of Berkeley mathematician Hung-Hsi Wu described the division between concepts and skills as a bogus dichotomy:[18]

> This bogus dichotomy would seem to arise from a common misconception of mathematics held by a segment of the public and the education community: that the demand for precision and fluency in the execution of basic skills in school mathematics runs counter to the acquisition of conceptual understanding. The truth is that in mathematics, skills and understanding are completely intertwined. In most cases, the precision and fluency in the execution of the skills are the requisite vehicles to convey the conceptual understanding. There is not "conceptual understanding" and "problem-solving skill" on the one hand and "basic skills" on the other. Nor can one acquire the former without the latter.

So, it is not surprising that countries with excellent mathematics education place great emphasis on skill development. Moreover, their approach to this is consonant with the recommendations of laboratory research. For example, skill practice should be *spaced,* not *massed.*[19] This means that instead of practicing the same skill in a small number of long stretches—such as doing a whole single-skill worksheet in one sitting, or spending half an hour on a single skill in class—skills should be practiced in a large number of small installments, such as five- or ten-minute daily practice sessions covering a wide range of different skills. A great number and variety of studies have found that spaced practice leads to better learning and retention of skills, even putting aside the obvious greater appeal it holds for students.

Furthermore, in most cases skills are likely best practiced only *after* the underlying concept is introduced and explained.[20] For example, before practicing adding fractions with the same denominator, students should learn *why* the procedure is what it is. This way, skills practice is not a meaningless ritual, but just the automation of a meaningful, sensible procedure.

Problem-Solving

One of the great thrills of mathematics—whether at the research level, in college, or in third grade—is the "aha!" moment. This is the instant when you are suddenly struck by the insight that unlocks the solution to the problem. It doesn't happen with routine exercises, but only with problems which are *new*: which you've never seen before, which have some complication or twist that seems to get in the way.

Take, for example, the following problem we could give to a fourth grader:

> A dragon had a treasure of gold coins. It lost twelve. Then half of the remaining coins were stolen by an enterprising hobbit. Following this, the dragon spent another five coins to change the locks. It was left with 35 coins. How many coins did it have at the beginning?

To solve this problem, the only mathematical topics needed are basic arithmetic operations with numbers less than 100. But we also need a clever insight: that we can work backward, starting with 35 and doing the opposite of each step until we reach the beginning.

This problem-solving strategy—working backward—is one out of dozens, hundreds, and thousands, each with innumerable variants. Is it important to incorporate such problem-solving into mathematics education? From a humanistic standpoint, the answer is a resounding *yes*: the nations with the best mathematics education do it, educational critics recommend it, and mathematicians consider it the very heart of their discipline. A mathematics course in which students constantly solve novel—even if only by being *slightly* different from the familiar—problems, feeling the joy of the "aha!" and the inventiveness of their own minds, is an immeasurably more aesthetically rich experience than a course full of the routine and predictable. And many mathematics teachers and educational critics have argued that it develops the general ability to think. As Paul Sally—who was my freshman year mathematics professor at the University of Chicago, and more importantly an influential figure in American mathematics education—put it, "Mathematics is a discipline of the mind that exceeds almost any other discipline of the mind. If you want to be able to think, mathematics is the way to do it."

Scientific research is conflicted on this question, and it is one of the great controversies of psychology: Is there really such a thing as "developing general thinking skills"?

In the early twentieth century, there was a heated national debate concerning the relative merit of academic and professional training.[21] Many of the time's educational theorists questioned the value of the traditional curriculum, which taught the classical subjects (including Latin) to students regardless of their likely subsequent careers. When they criticized these subjects as irrelevant, the defenders of the traditional curriculum replied with what became known as the *theory of formal discipline*: that subjects like mathematics and Latin strengthened the mind, developing general thinking (or, in the case of Latin, linguistic) skills. In one of the most influential psychology experiments of all time, Columbia University, Teachers College professor Edward Thorndike* (1874–1949) put this theory to the test. Thorndike taught one group of students to measure angles with a protractor, while another (the control group) was not given this lesson. Then, he gave both groups a lesson on measuring line segments with a ruler. If the theory of formal discipline was true, Thorndike argued, then the group schooled in protractors should be better equipped to learn about rulers: the two skills are, after all, quite close. But the experiment revealed no difference in performance, leading Thorndike to reject the theory of formal discipline.

There were other experiments with the same essential finding, including one which found that a year of Latin made no discernible

* Thorndike got his start in animal behavior, conducting experiments on cats' ability to go through mazes. He was widely regarded as a world expert on cat navigation.

impact on subsequent English language performance. In modern psychology parlance, this was evidence that learning did not *transfer* beyond the narrow context in which it first took place. Many psychologists and education reforms of the time concluded that talk of "general thinking skills" was nonsense, that advocates of classical education were trapped in the past, and that the only responsible education should aim to impart very specific skills that will be of use to students in their careers. There are dark undertones to this line of thinking: if general intelligence cannot be improved through good education, then what is left except to sort students into different tracks early on, training each for their manifest social and professional destiny? Some proponents of this view—including Thorndike himself—went so far as to endorse eugenics.

Fortunately, subsequent research has painted a more nuanced picture. While transfer of learning does not happen as readily as one might naively think, it nevertheless does occur when the conditions are right. For example, students who are taught how to solve one problem of a certain type might not transfer their learning to a new one, but if shown *two* problems of that type the students *will* successfully solve the new problem.[22] In general, knowledge of a certain idea is more likely to transfer when it is *deeper*—i.e., more suffused with genuine understanding—and when students have seen a wider range of examples and applications of the idea.[23]

Taking all of this into account, the modern perspective is that *transfer is hard*. Not impossible, but hard. And the *further* the transfer—the more different the new task from the initial knowledge—the harder. Because of this, today's psychologists are generally quite skeptical of claims that some or another subject improves general thinking skills: if obtaining even *near* transfer—such as to a similar mathematical problem—is hard enough, what then is to be said of the massively far transfer of "general thinking skills"? There is cause to doubt it, and no experimental evidence to support it.

On the other hand, there is apparently conflicting evidence from another field of research. Scientists who study the relationship between genetics ("nature"), environmental influences ("nurture"), and general intelligence (as measured, for example, by IQ tests) have found that nature and nurture play a roughly equal role.[24] For example, identical twins raised in the same environment usually end up with quite similar IQs, but identical twins raised by different families can end up with considerable IQ differences. If the environment plays such an important role, how could schooling—which can account for half of children's waking hours—not matter, too? Some studies have even found evidence for a direct causal link between more schooling and higher IQ.[25]

What explains the apparent discrepancy in the evidence from these different sources? There is no easy answer, but it may have to do with the slow pace of genuine education. If students do develop general thinking skills, this happens not through quick fixes—like the homeopathic dose of Latin afforded by a one-year course—but through a years-long effort of systematic instruction, with each month and year adding a drop to what will one day become, if not an ocean, at least a respectable pond of intellectual ability. Laboratory experiments, which are constrained to short spans of time and narrow subject matter, can never hope to measure such effects.

Returning then to our original question—namely, whether teaching genuine mathematical problem-solving is important—there is no obvious answer. Problem-solving may develop general intellectual skill, but whether this is true—and even whether anything at all develops such skill—remains the subject of some controversy. This leaves each person to decide on the basis of values. Those who see education as a narrow undertaking, serving to equip students with concrete skills for daily life and work, will see little value in teaching the sort of problem-solving that requires insight and invention. Meanwhile, those who see education as an aesthetic and cultural experience, aiming to initiate students into the world of the mind, will consider it to be an indispensable part of good mathematics instruction. My own sympathies are decidedly with the latter: I feel that it is a crime against both the student and the subject to turn mathematics into a lifeless sequence of routine exercises, stripping away its natural beauty and joy. But others no worse informed may legitimately disagree, and I will resist the temptation to claim that such people must hate children.

On the separate question of *how* to best teach mathematical problem-solving, we are back in the comfortable territory of clear agreement between experimental and humanistic evidence. Successful problem-solving requires *metacognition*: purposeful thinking about one's own reasoning process.[26] For example, students must learn to make careful sense of a problem before trying to solve it—a seemingly obvious step to experienced problem-solvers, but actually not one students know to take on their own. Psychologists studying metacognition have found that it is developed through exposure to a wide variety of examples. Teachers can demonstrate their thinking process to students by solving a problem "out loud" in front of the class; furthermore, teachers can provide coaching to students, guiding them to using the right thinking strategies as they work on problems. After a certain way of thinking has been demonstrated and practiced hundreds—or perhaps thousands—of times in a wide variety of different problems, it will start to sink in; after some months or years, it will become a part of the student's own metacognitive arsenal.

Practices entirely in line with the recommendations of research are to be found in nations with strong mathematics education traditions, and in the writings of leading educational thinkers. The best-known example is *How to Solve It,* an international bestseller—if a book about mathematics can be such a thing—written by Hungarian-American mathematician George Polya, who spent most of his career at Stanford University.[27] In this book, Polya describes a method for teaching problem-solving skills; while the method is not of his own invention—but rather something observed widely among excellent mathematics teachers—the book was the first to systematically describe it.

Per Polya's method, the teacher will work problems in front of the class, demonstrating ways of thinking through asking questions. For example, the teacher will ask, "What are we trying to find?" This carries the implicit strategy: begin every problem by clarifying the goal. The teacher asks: "What information are we given?" This means: always understand the givens. The teacher asks: "Is there anything that, if we knew, could allow us to get the answer in one step?" This means: consider working backward. After students have heard these questions repeated a thousand times—and in a thousand completely different problem situations—they will absorb them. Now, when a student is solving a problem independently, a voice in their head will ask: "What are we trying to find?" And the student will know what to do next.

A key point is that these strategies are not directly taught to students. A teacher could certainly pull students aside and hand them a list of strategies, but this would have little benefit. To acquire fluent, flexible mastery of the strategies, students must internalize them: they must become second nature, automatic ways of thinking about any problem. And this can only be done through repeated exposure and practice in a very wide range of different contexts.

From this, we can see that excellent problem-solving instruction requires three ingredients. First, the correct method must be used to ensure that students truly learn strategies (instead of superficially memorizing them). Second, each strategy should be practiced throughout the course on a very wide variety of different problem types; the wider the variety, the more likely students will be able to apply the strategy to new problem types—in other words, the further they will be able to transfer the knowledge of the strategy. And third, a wide range of strategies should be taught: not just a handful, but a full collection that will allow students to tackle almost any unfamiliar problem appropriate to their grade level. In other words, the goal of good problem-solving instruction is not to memorize routine procedures for solving a handful of familiar problem types, but rather to develop *richness and flexibility of thinking* when working on mathematical problems of *any* kind.

Rigor

Professional mathematicians often speak of "mathematical rigor." By this they do not mean *difficulty*, but rather *precision*. A statement is rigorous if it means exactly what it says, and says exactly what it means. The opposite of "rigorous" is not "easy"—it is "vague," or perhaps even "sloppy."

For example, if asked what a decimal is, a student might say: "A decimal is a number that also has tenths and hundredths after the point." This definition is lacking in many ways; it is easy to find examples of decimals that don't fit it, and also things that fit the definition but are not decimals. For example, 2.1 doesn't fit because it lacks hundredths, and the definition makes it seem as if these might be essential; 2.145 doesn't seem to fit because it has thousandths, but the definition leaves no provision for these; π.24, by which we mean

$\pi + 0.24$, appears to fit—even though it is at best an unusual way of writing a number. The sloppy definition neither means what it says nor says what it means. We would say that it lacks rigor.

By contrast, an improved definition might be this: "A decimal is a mixed number whose fractional part has a denominator which is a power of 10. We write decimals by placing the whole part first, then a period, and then the numerator of the fractional part." This solves the problems that the previous definition had. However, it too has a flaw: taken literally, it would have us write 2 5/100 as 2.5. To fix this, we would need a further sentence which clarifies that in cases when the numerator has fewer digits than one less than the denominator, we should remedy this by supplying extra zeros in our notation.

As this example illustrates, increased rigor can sometimes lead to longer, more complicated statements. In other words, a quest for perfect rigor would force us to sometimes sacrifice clarity. Given this difficulty, is rigor worth pursuing in school mathematics? In research mathematics, the case for rigor is apparent: rigorous statements allow professional mathematicians to be sure that they are not making mistakes in their reasoning. But in school mathematics, isn't clarity more important? And isn't rigor irrelevant?

This is not a question that easily lends itself to scientific study, and I am not familiar with any laboratory research on the subject. There is, however, a compelling humanistic case for the importance of rigor in mathematics education. To begin with, rigor supports clarity in the majority of cases: precise statements are harder to misinterpret.

But there are also two deeper reasons to seek rigor, and these are similar to the humanistic arguments for the value of problem-solving, as discussed in the preceding section. The first reason is that rigor is an essential aspect of mathematics as a discipline. The perfection of high rigor is part of mathematical aesthetics, and no one can say that they understand mathematics if they do not have a sense for it. Thus, if we wish to truly initiate students into the discipline, we must teach them the meaning of mathematical rigor—and this can only be taught by making it an essential part of how mathematics is done in every class. Adults cannot teach children honesty through lectures—they must teach it through constant example; it is the same with mathematical rigor, which is a form of intellectual honesty.

The second reason is that we might hope that training in mathematical rigor will transfer to other aspects of students' intellectual lives. A student who has learned to say what they mean and mean what they say in mathematics would be well-prepared to do the same in history, literature, and science. Moreover, she would be equipped to carefully dissect the statements of teachers, writers, colleagues, and politicians, telling apart the sense from the nonsense—and, perhaps most importantly, from the almost-but-not-quite-sense. As with problem-solving skills, this is a wish for "far transfer"—the hope that training in mathematical rigor will give students the sort of general critical thinking skills that will translate to all aspects of their lives; and as with that example, it is neither supported nor conclusively rejected by laboratory research. We cannot have certainty that it is valid. In my personal experience, I feel that it is true: I believe that my mathematical training has made me a far more meticulous reader of all manner of texts than I was before. But my personal impression is, of course, just that—and not to be taken for conclusive proof of anything.

As an additional piece of evidence in support of rigor in school mathematics, the cool kids—by which I mean the nations with the strongest mathematics education systems—do it. This is apparent in widely used textbooks in countries like Singapore, Russia, China, and Japan, just as the lack of rigor is apparent in most of the textbooks used in American schools.[28] For countries with the strongest mathematics education, rigor is a deliberate choice. For example, the Russian tradition owes it to the great Leonhard Euler,[29] who spent most of his career in St. Petersburg at the invitation of the Russian empress, Catherine the Great. Besides possibly being the only time in history that an empress sought the company of a mathematician, Euler's time in Russia permanently shaped the nation's mathematics education system. He wrote school textbooks and trained curriculum developers. And he instituted an important principle: school mathematics should balance rigor and clarity, seeking to achieve as much as possible of each.

Applying the Principles in Practice: The Paramount Importance of Trial and Error

The practices discussed in the preceding several sections may succeed in separating good from bad mathematics instruction: the nations with the best instruction generally adhere to them, while the rest are usually lacking in one or more. Even then, good mathematics instruction is not as simple as marking off a handful of items on a checklist. The qualities I have identified—appropriate instructional guidance, a focus on understanding, adequate development of skills, an emphasis on mathematical problem-solving, a good dose of rigor without loss of clarity—are very broad, and can be present in varying degrees on each page of a textbook and in each segment of a lesson. Delivering instruction that exhibits these qualities requires more than just good books: it requires the ability and the constant, unrelenting effort of teachers to inject them into every minute of instruction.

To put this in context, a good school mathematics textbook might be a few hundred pages long. But if you wrote down everything a teacher said, wrote, and did in class, it would easily run to fifty pages per lesson—adding up to thousands of pages for a course. Thus, textbooks are only the tip of the instructional iceberg. Nine out of ten pieces of instructional content come not from the book, but directly from the teacher.

What is required of teachers to ensure each lesson adheres to the principles of good mathematics instruction? A very great deal. Approaching each topic in a manner that supports genuine understanding is difficult, and requires that the teacher knows the material deeply. Furthermore, the teacher must know which ways of explaining the material are most likely to resonate with children. The development of skills is challenging, too: textbooks rarely prescribe the extent of skills practice needed—which varies by skill—or at what intervals it should take place. Successfully teaching problem-solving means choosing a wide variety of excellent problems, but also it means knowing how to model the process of attacking each one. This is different for every topic, and moreover for every problem within a topic. Rigor is not just important in textbook passages: the teacher must speak in a mathematically rigorous manner throughout every lesson, and must be able to recognize when students do this. In other words, there are a thousand difficult instructional decisions that must be made in each lesson, beyond what is offered in textbooks. *The quality of instruction resides in the details.* The importance of this cannot be overstated.

And *how* can the quality of the details be assured? What is needed of the teacher? Certainly, knowledge of mathematics. Certainly also a familiarity with the textbook. But beyond this, a great deal of knowledge is needed that can only be obtained through *trial and error.* For example, how can you know for sure if a certain explanation will make sense to children? How can you know when a certain skill will be forgotten, and in need of review? How can you know when a certain formulation is *too* rigorous at the expense of clarity, and thus of no use to students? How can you know which problems will prove too challenging, and which too routine? We may make educated guesses at these things, but even the most astute guesser would likely find himself often in the wrong. Definitive answers to these questions are only obtained through experience: in other words, through trial and error. This is one reason why there are so few exceptional traditions of mathematics education; few countries have had the opportunity to experiment for decades, systematically learning from their experiences and passing them on to the next generation. In Japan, this has been done through *lesson study,* the Japanese tradition in which teams of teachers experiment with a chosen lesson and then publish their findings.[30] In Russia, an important role was played by so-called "methodologists," whose charge is to understand past curricula and to assist in the development of new ones.

An illuminating analogy is to the *usability testing* of software. Interface designers can apply cognitive psychology, can rely on established design conventions, and can employ their judgment acquired from their past experience. But none of these things *guarantee* a successful interface: usability is so idiosyncratic that a failure can occur in completely unexpected ways. For this reason, interface designers worship at the altar of *usability testing.*[31] An hour with five users reveals more than a decade of speculation. Mathematics instruction is the same: it requires a bedrock of strong mathematical content knowledge, but beyond this only experience can yield the knowledge required for true excellence. And, just as in the case of software interfaces, much of this experiential knowledge is not of a general nature: it is particular to each moment of every lesson. So we have this: *instruction that is excellent in all of the details can only be achieved through extensive trial and error.*

There is another important point to be made about excellent mathematics instruction. I have spoken of *details,* but these details are not isolated from one another. The pieces of a mathematics course are interrelated, and must hang together. The most obvious example is terminological consistency: if we speak of "natural numbers" now, "whole numbers" to mean the same thing a minute later, and "positive integers" synonymously immediately afterward, students will be confused. But there are more subtle ways in which a course must be a cohesive whole. For students to develop good problem-solving skills, they must be given problems that gradually and systematically develop these skills; for example, a certain strategy (say, making a table) can be taught once, then practiced in the next week with some simple problems, then applied in the subsequent month in some new contexts, and then polished over the following year with a wide range of problems of varying difficulty. For students to learn with genuine understanding, each topic should be presented in a way that builds on the preceding ones. This means not only that students need to have *covered* prerequisite topics, but also that the *way* in which they were covered ties in with the way the new topic is presented. For example, if fractions were first introduced using a pizza cut into equal slices, then this same model will be a natural choice for introducing the addition of fractions—and not another model, such as four quarters making up a dollar.

There are many other examples, but these should be enough to see the general point: *good mathematics instruction demands close coordination between the details of all of the lessons making up a course*. The term most widely used for this notion is "coherence."[32] As our examples illustrated, coherence is not an end in itself, but rather an essential ingredient for a course to follow the principles of good mathematics instruction. An incoherent course cannot successfully teach problem-solving, nor can it engender genuine understanding. In fact, this is a special case of something we discussed in the preceding chapter: the systemic nature of instruction. Then, we discussed mainly how the administrative pieces of an education system—such as teacher preparation and the structure of schools—must hang together, but now we see that the same principle applies to the innards of a mathematics course.

Together, these three facts—that quality resides in the details, that trial and error is essential, and that a course must be coherent—account for much of the difficulty in designing and delivering excellent instruction. They have important and far-reaching consequences: they imply that good textbooks are necessary but far from sufficient for good instruction, that designing good instruction from scratch on the first attempt (or even the second, third, or fourth) is virtually impossible, and that outstanding courses must be made in toto rather than being assembled from spare parts. As we will see in later chapters, these consequences will offer a strong motivation for the practice of instruction modeling.

The Supports: Textbooks, Lesson Plans, Teacher Preparation, and Advanced Programs

By now it is clear just how *difficult* it is to provide exceptional instruction. So it is not surprising that countries with outstanding mathematics education have a host of supports to make it possible.

First come excellent textbooks: books that explain the ideas of the course, allowing the reader to truly *understand* them; that include good exercises for developing basic mathematical skills; that offer a generous variety of well-chosen problems, appropriately sequenced to support the learning of general problem-solving strategies; and that have the virtues of rigor and clarity. These textbooks serve students, but—even more importantly—they serve teachers, setting the tone for the course and furnishing the main material for classroom lessons and assignments. They are usually concise—two or three hundred pages—and printed in at most two colors, with sparing illustrations to avoid distracting from the substance.[33]

Second comes access to good lesson plans to go with the textbooks. Sometimes—as in Russia—these plans are developed by the authors of the textbooks; other times—as in Japan—they are developed and published by independent teams of teachers.[34] But in all cases, they have been carefully designed and field tested to ensure quality. A good collection of lesson plans ensures that time is spent efficiently and that all of the necessary components are present: explanations of all relevant concepts, skills practice, problem-solving, guided work, independent work, and everything else needed for complete mathematics instruction. Teachers can certainly modify lesson plans to make them their own, and more experienced teachers eventually develop their own plans from scratch, but they begin their careers with a ready-made set of high-quality plans developed previously.

Third comes strong teacher preparation in mathematics. This preparation can happen before teachers enter the classroom—for example, during their college training—or afterward, but usually there are components of both, and always there is a strong emphasis on deep understanding of mathematics.[35] Russian secondary school teachers take five years of college-level courses, mostly in mathematics and mathematics pedagogy; Japanese teachers spend hours each week studying lesson plans in groups with their colleagues; Chinese teachers undergo a years-long apprenticeship with their more experienced colleagues. These intense training regimens are not surprising: beginning in the late 1980s, the University of Michigan's Deborah Ball and her colleagues carried out now-famous work to systematically identify the knowledge needed for excellent mathematics instruction, and found that a massive quantity of very specialized knowledge is required.[36] This body of knowledge—collectively called "mathematical knowledge for teaching"—includes knowledge of mathematics itself, knowledge of how students will respond to this or the other explanation, knowledge of typical student misconceptions and how to address them, knowledge of connections between mathematical ideas, and much else besides.

The fourth key support for excellent mathematics instruction is—perhaps surprisingly—the presence of strong programs for mathematically advanced children. Far from serving only the needs of the participating students, these programs benefit the nation's education system as a whole. Good programs for children who are especially advanced or motivated in mathematics are a pipeline for future mathematics teachers; their participants' enthusiasm for the subject rubs off on their peers; and they help to attract talented individuals to mathematics teaching—who often begin by working with advanced students, and develop an interest in general education only later. These programs include specialized mathematics schools, summer mathematics camps, after-school programs, and programs designed to prepare students for mathematics competitions.

This serves as yet another illustration of the *systemic* nature of education. Good mathematics instruction requires a large collection of things to be done well, and these things are interconnected. Textbooks, lesson plans, and teacher preparation must all be linked: the lesson plans must be designed around the specific textbooks, and teacher preparation must address both. If the components look good in isolation but are not designed to work together, the result will not be sufficient to enable outstanding mathematics instruction.

Where the United States Stands in Mathematics Education

How well do students in the United States learn mathematics? On international comparison tests, American students are roughly average among developed nations. However, this ignores the vast variance within the nation: students in the highest-scoring states, such as Massachusetts, match the leading nations in performance, while students in the lowest-performing states are closer to nations from the developing world, such as Armenia and Lebanon.[37]

But as we saw in the first chapter, these scores reflect a great many influences besides instruction. And the question we *really* want to answer is about the nature of instruction, since this is something we have more control over than cultural, social, or economic circumstances. This is a reasonable question to ask, since—as Stigler and Hiebert argue in *The Teaching Gap*—instruction is fairly consistent across the different classrooms within a country:[38]

> Readers who are parents will know that there are differences among American teachers; they might even have fought to move their child from one teacher's class into another teacher's class. Our point is that these differences, which appear so large within our culture, are dwarfed by the gap in general methods of teaching that exist across cultures.

And how is the quality of mathematics instruction in the United States? To get an idea, we can analyze it in terms of the various attributes of good mathematics instruction discussed in this chapter.

In the matter of the degree of guidance, American classrooms are predominantly teacher-led: Stigler and Hiebert report that only 9 percent of tasks in American mathematics lessons were predominantly controlled by students, compared to 40 percent in Japan.[39] But as we saw in the first section of this chapter, this in itself is not necessarily a negative quality. It is more important to ask whether American students are taught to understand the mathematics they learn, receive appropriate skills practice, are taught mathematical problem-solving skills, and develop an understanding of mathematical rigor.

In the matter of *understanding*, American mathematics education has room to improve. For example, in *Knowing and Teaching Elementary Mathematics*, Liping Ma compares how American and

Chinese teachers handle column subtraction.[40] Procedurally, to compute 75 − 12, we subtract each place separately, obtaining 63. The reason why we subtract in this manner is that 75 is 70 + 5, and 12 is 10 + 2. So we can subtract the tens and ones separately and then combine the results. In cases like 21 − 9, we must borrow: since 9 is greater than 1, we "borrow" 1 from the tens' place; then we subtract 9 from 11, and obtain a final answer of 12. Conceptually, this works for the same reason as the previous example: 21 is 20 + 1, and therefore we can regroup it as 10 + 11 and proceed as before.

Column subtraction can be done either with or without understanding. It is entirely possible to go from digit to digit, "borrowing" as needed, without any notion of ones, tens, groupings, and the like. Ma found that 83% of the American teachers in her study understood and taught subtraction primarily on this procedural level. By contrast, only 14% of the Chinese teachers did the same: the majority of the Chinese teachers emphasized the *reasons* behind the procedure, and 35% even went so far as to teach multiple ways of regrouping.

As this example illustrates, students in the United States are usually taught mathematics as a collection of arbitrary procedures to be followed without questioning. The late Paul Sally, my freshman year college mathematics professor, mocked the prevailing approach to teaching division of fractions with verses adapted from Tennyson's *The Charge of the Light Brigade*:

Ours is not to reason why,
Just invert and multiply!

When this is the approach, it is no wonder that many in the United States conclude that they "cannot do mathematics." When the subject is taught as a random collection of meaningless rituals, no one can do mathematics.

In the matter of skills practice, American students have plenty of it. This is a positive side of American mathematics education, as skills are quite important. However, there are two caveats. First, skills that are not grounded in conceptual understanding are harder to remember—after all, it is easier to remember a meaningful procedure than one whose steps seem completely arbitrary. And second, American students usually practice mathematical skills in long, homogeneous stretches, completing lengthy worksheets with dozens of nearly-identical exercises. This form of practice—*massed practice*—has been shown by many psychology studies to be less efficient than frequent practice in small installments, known as *spaced practice*. Thus, while considerable time is devoted to skills practice, this time is not spent in the best possible way.

Turning next to problem-solving, we again find deficiencies. Good problem-solving instruction calls for a rich array of different problems, requiring many different problem-solving strategies, and each problem having its own twist. In the United States, this variety is lacking. Students do plenty of basic exercises (adding numbers, solving simple equations, and so on), and also word problems meant to appear "realistic"—such as using a table to determine how many muffins Mr. Zambini sold in his store on Wednesday. But the challenge in such word problems is usually reading comprehension, as the underlying mathematical logic is usually simple and familiar. And partly due to the emphasis placed on end-of-year state testing, teachers avoid imaginative problems, preferring to focus on standard problem types that will appear on the exam.

In the matter of mathematical rigor, this notion is almost entirely absent in American mathematics education. Textbooks do not use careful phrasing, and virtually every page of the most popular textbooks contains statements a mathematician would consider false.[41] Teachers follow the example set by the textbooks, rarely attending to correct mathematical speech and writing. As a typical example, I once sat in a seventh grade lesson where the teacher kept speaking of the "north" and "south." This was not a history class about the Civil War: it was a mathematics class, and the words were used to mean "numerator" and "denominator."

This avoidance of terminology is only one facet of the problem. For example, if a student in a Russian mathematics class said, "To add fractions, we add the numerators," the teacher would correct the student's phrasing, accepting only something along the lines of, "To add fractions with equal denominators, we add the numerators and keep the denominator the same." In the United States, by contrast, teachers almost always accept any answer when the intent was correct—even if the phrasing itself was vague or even outright wrong.

Given these deficiencies, we would expect that the supports needed for good mathematics instruction—good textbooks, lesson plans, teacher preparation, and advanced programs—would be deficient as well. Indeed, the quality of mainstream mathematics textbooks in the United States is exceedingly low. The books are needlessly long—often over 700 large pages—and printed in garish colors. They attempt to catch students' interest with anything but mathematics: you will find photographs of rollercoasters, sidebars about mathematical careers, illustrations of seashells (presumably, because their spirals appear "geometric" and therefore are thought by the book authors to be germane to the subject), and myriad other attempts to dress up what the authors must consider a singularly boring subject. What you will *not* find is precise mathematical language, clear discussions of mathematical ideas that explain the reasons for things, or interesting mathematical problems. Berkeley mathematician and leading educational critic Hung-Hsi Wu coined the term "textbook school mathematics" (abbreviated TSM) to refer to the discipline represented in these books.[42] In other words, to mathematicians, what appears in American school textbooks is not recognizable as mathematics.

The situation with lesson plans is similar. American teachers rarely have access to good lesson plans for courses, instead being left to fend for themselves. Almost all of the American mathematics teachers I know had to develop their own lesson plans from scratch during their very first year of teaching; at most, they had access to lesson plans developed by other teachers in their school. Even if individual schools had the expertise to develop consistently excellent lesson plans—which is unlikely, since in countries like Japan the development of lesson plans is seen as a nationwide effort—they would be stymied by the constant changes in textbooks and state content standards.

As concerns teacher preparation, American mathematics teachers are not given the same training as their counterparts in countries with the best mathematics instruction.[43] The curricula of American teaching colleges focus on general teaching strategies, education history, policy, and sociology, but rarely provide the courses needed to give their graduates a deep knowledge of school mathematics. There are some exceptions: for example, the UTeach program at the University of Texas offers coursework to prepare mathematics and science majors for careers as teachers, and has been replicated in dozens of other universities with considerable success. However, such programs only contribute trace amounts to vast national cohorts of newly minted teachers.

Another source of new teachers are *alternative certification programs*. These programs—which allow those without teaching degrees to enter the profession—include Teach for America, which places elite recent college graduates in inner-city schools. Teach for America has done remarkable work to bring thousands of talented people into education; many leave after a few years, but many others remain. However, this does not solve the problem of mathematical content preparation. And furthermore, like UTeach, Teach for America accounts for a microscopic fraction of teacher production. The majority of alternative certification programs are diploma mills, helping to meet teacher shortages by lowering the standards of admission to the teaching profession.[44] One large Texas program advertises throughout the state on billboards which ask, "Want to teach? When can you start?" We should be glad there is no sister program for surgeons.

The analogy with medicine is a revealing one. We regard medical doctors as generally capable individuals, turned into medical experts by years of systematic training. Given the choice between an ordinary individual who has gone through the rigors of medical school, and an unschooled but—we are assured—"natural medical genius," we would much prefer to go under the knife of the former. Yet in teaching, everything is reversed. In the mind of many Americans, great teachers are born, not made. We imagine the charismatic teacher, bringing her passion and energy to bear on the problem of transforming young minds; the students resist at first, but ultimately succumb to the giant's sheer force of personality; the students graduate, against all odds, and enroll in elite colleges; touching scenes of proud parents; the credits roll, applause from the audience.

I am not suggesting that charisma and energy have no importance to a teacher—the contrary is obviously true. However, by elevating these qualities above all else, we unwittingly undermine teaching as a profession. Any profession, after all, is defined by its professional body of knowledge—a large set of norms and competencies that allow the profession's practitioners to preform consistently. Good professionals—be they doctors, lawyers, or engineers—are defined above all else not by standing out, but by fitting in. "Failure to conform to professional norms" is the highest censure of any profession. In medicine, it is called malpractice.

American society is quick to blame teachers as a group. Many popular government policies—such as systems that punish and reward teachers for student performance—carry the implicit belief that perceived educational shortcomings can be remedied, if only teachers as a group can be led to reason. Ironically, this willingness to blame teachers and our society's worship of hero-teachers are different sides of the same coin: both are caused by our attribution of good teaching to the individual teacher, rather than to the supports that we as a society have made available. Yet international comparison studies show that the supports are what makes the difference.[45] Instead of blaming teachers, we should provide better teacher education, giving teachers the body of knowledge needed to provide excellent mathematics instruction.

In the matter of advanced mathematics education, the picture is more positive. There is a fair number of outstanding summer programs for children with special interest in mathematics—for example, I was converted to the discipline after attending a program called Canada/USA Mathcamp. A New York City program called Bridge to Enter Advanced Mathematics provides an analog of Mathcamp for middle school students from poor schools; the program has been such a success that many alumni have gained admission to the city's most competitive high schools. There is a burgeoning "math circle" movement: based on Russian after-school mathematical enrichment clubs, math circles have been established in dozens of American cities. A handful of schools—including New York's Stuyvesant High School, Virginia's Thomas Jefferson High School for Science and Technology, and the Illinois Mathematics and Science Academy—offer exceptional mathematics instruction for advanced students; a new school in San Francisco, called Proof School, promises—based on the individuals involved, who are among the leading figures in advanced mathematics education—to become one of the best specialized mathematics schools not just in the country, but in the world. The United States routinely places in the top three countries in the International Mathematics Olympiad, the world's foremost high school mathematics competition. And finally, many of the world's best research mathematicians are products of American school education.

Nevertheless, there are some concerns. The best programs still only serve a tiny fraction of the nation's mathematically advanced students, and all but a handful are de facto only available to students from wealthier families. There is very limited interaction between the country's advanced and general mathematics education systems: few people take part in both, either as teachers or as developers of instructional materials. And many of those involved—including many of the coaches and team members of the International Mathematics Olympiad teams—are first- or second-generation immigrants, raising questions about the future sustainability of the programs.

To summarize, the United States has strong programs for some advanced students, but is far from outstanding when it comes to general mathematics instruction. This may seem like a bleak assessment, but it is important to remember that it is a comparison against *ideal* mathematics instruction, not against *average* instruction. The vast majority of nations—including developed ones—are in the same situation, and only a handful would score high on all counts.

Many education advocates call for urgent reform, predicting imminent national demise if nothing changes. Personally, I find it difficult to agree. First, American education as a whole has many positive qualities: American teachers are dedicated and hard-working, schools do a great deal to develop students' leadership and communication skills, and parents often arrange a rich array of extracurricular activities.* Comparisons with other countries often ignore these things. And second, mathematics instruction in the United States was not considerably different in past generations—and, in fact, in the past it was likely somewhat worse[46]—but this has not prevented the nation from producing or importing outstanding mathematicians, scientists, and engineers. Panic is not called for, and in fact could be counterproductive, leading us to seek quick fixes instead of the sustained effort the task requires.

* That said, these strengths of the American educational system are very unevenly distributed: minority students and those from poor families receive a substantially worse education than their privileged peers. For a detailed discussion of this serious problem with equity in American schools, see Linda Darling-Hammond's book *The Flat World and Education: How America's Commitment to Equity Will Determine Our Future*.

But while I do not endorse panic, why accept the current—adequate but uninspiring—state of mathematics education in our nation? By striving toward the ideal, we can strengthen our production of scientists and engineers, expand children's career options, and reduce educational gaps between the rich and poor. But just as importantly, we can give students a taste of what mathematics is truly about: not a tedious course of thoughtless mimicry, but a beautiful discipline of the mind. As students truly *understand* mathematical ideas, experience "aha!" moments as they solve unconventional problems, and learn to speak and reason rigorously, they will come to know the joys of what makes us distinctly human: our capacity for thought and our thirst—as Vergil put it—to understand the reasons for things.

Chapter 3

Instruction Modeling
Replicating the Practices of Expert Teachers

If only a handful of countries have truly outstanding mathematics instruction, and if we can find out which ones, why not bring their methods over to our own country? Indeed, curricula have at times crossed national borders. The Singaporean program was written by authors educated in the Chinese tradition,[1] Russia adopted German programs in the eighteenth century, and many Eastern European countries shared textbooks and greatly influenced one another's education systems.[2]

However, importing one country's education system into another is far from easy. As the previous chapter makes clear, providing good instruction in mathematics—and in any subject, for that matter—is difficult. And it is especially difficult to provide good instruction on the level of an entire nation. After all, it requires a large number of supports to be in place, and these supports must be designed to work well together as a system. This means that introducing, for example, the Japanese approach would require importing Japanese textbooks, Japanese lesson plans, Japanese teacher colleges, Japanese ways of

Instruction Modeling. George A. Khachatryan, Oxford University Press (2020)

DOI: 10.1093/oso/9780190910709.001.0001

structuring the school day to ensure professional learning opportunities for teachers, and much else besides. To do this properly, all of the specialized knowledge of Japanese teachers and university education instructors would need to be transferred to the minds of their analogs in the United States.[3] Can you imagine the size of this knowledge base? Even if it were neatly laid out somewhere in a set of books—and it most certainly is not—it would surely make up tens of thousands of pages. This makes instructional reform difficult even when a single agency controls all of the components of a country's education system; in a country like the United States, where states and school districts are largely independent of one another, it is close to impossible.

This has not prevented many from trying. While there have been some successes, failure has been the more common outcome. One colorful example is related by Berkeley mathematician Alexander Givental, who tells a story of an attempted implementation of Singaporean mathematics education in the United States.[4] Instead of adopting the approach as a whole, schools increasingly latched on to individual features of the Singapore curriculum, thinking that this would be enough. In particular, Singaporean textbooks taught students to solve certain word problems by drawing so-called bar diagrams. For example, if asked to find how many apples Mary had if Rosie had five and together they had twelve, students would draw something like Figure 3.1.

From this it would be clear that the answer could be found by subtracting five from twelve. Givental recounts how bar diagrams increasingly came to dominate views of Singapore mathematics: they became the central topic of workshops, the subject of wall posters, and the focus of books. Teachers believed that as long as they taught students to solve word problems using bar diagrams, they were following Singapore. It seemed that no one realized that the

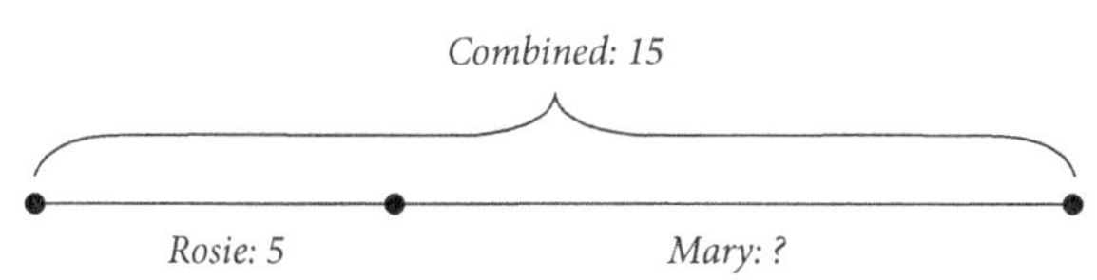

FIGURE 3.1 A diagram of the sort used in Singaporean arithmetic instruction.

bar diagrams—while indeed used in the Singapore textbooks—represented a tiny sliver of the total collection of ideas and methods that made up the course.

This is not the fault of the teachers and school administrators. Rather, it is a mistake that any person in their position would be apt to make. After all, Singaporean teachers do not learn their craft from a short textbook and a week-long training session: they learn it through *cultural immersion*. First they attend Singaporean schools as students, experiencing the methods firsthand. Then they spend years in Singaporean teacher colleges, learning from instructors already well versed in the curriculum. And finally, they find themselves in schools, working alongside their more experienced peers. Throughout, they have opportunities to observe and be guided by peers who already know the methods. Expecting American teachers to pick up the methods without a similar experience is like expecting someone to learn figure skating from a textbook: neither realistic nor humane.

Stigler and Hiebert refer to this problem as the "stability of systems of teaching." They write:[5]

> Because teaching is a complex system, these attempts to change it generally don't work. It has now been documented in several studies that teachers asked to change features of their teaching often modify the features to fit within their pre-existing system instead of changing the system itself. The system assimilates individual changes and swallows them up. Thus, although surface features appear to change, the fundamental nature of the instruction does not. When this happens, anticipated improvements in student learning fail to materialize and everyone wonders why.

In other words, instructional reform through conventional methods—like changing out the textbooks—is nearly impossible. The entire *system* of instruction needs to be changed, a herculean task. What then can be done?

Instruction modeling is aimed at precisely this challenge. It uses blended learning to systemically reform a variety of aspects of instruction all at once: the curriculum, the content of lessons, the problems, the explanations of ideas, and even the mathematical knowledge of teachers. In this chapter and the one that follows, we will find out how.

Expert Systems, the Inspiration

The inspiration for instruction modeling came from *expert systems*, one of the earliest forms of artificial intelligence.[6] The term "artificial intelligence" gained widespread acceptance at a seminal conference held at Dartmouth College in 1956.[7] By the middle of the twentieth century, computers could already accomplish routine, procedural tasks, such as multiplying numbers or sorting lists. The new field sought to build computer programs that could go beyond this, performing the complex, ambiguous tasks that we naturally assume only humans can perform, such as playing board games, recognizing handwriting, solving mathematical word problems, or driving a car.

Over the next several decades, computer scientists working on artificial intelligence developed many different approaches. Some approaches aimed to carry out human tasks without mimicking human ways of thinking: for example, today's websites that suggest television shows based on the visitor's viewing history employ sophisticated statistics—an entirely different method from what a friend would use in making the same suggestion. But other approaches went straight for the jugular, seeking to reason just as the mind does; it is in this second category that we find expert systems.

The first expert systems appeared in the 1970s, developed by Edward Feigenbaum's Stanford Heuristic Programming Project.[8] The main idea of expert systems was to study the thought process of human experts in a certain field, and then to build that same reasoning into a computer program. For example, in diagnosing a patient, an expert doctor might use the following* rules of deduction:

If the patient has a fever, check whether he or she has diarrhea.
If the patient has a fever and diarrhea, suspect food poisoning.
If the patient does not appear pale, rule out internal bleeding.

A computer scientist building an expert system would interview the doctor, extracting as many such rules as possible. At the end, there could be hundreds, thousands, or even millions of rules uncovered. The rules would be entered into the computer program, which would then "execute" the rules, just as a doctor would run them in her mind.

* These particular rules are completely made up. Please do not take as medical advice.

Whereas many previous forms of artificial intelligence had tried to create a machine that could reason like a human in any situation, expert systems changed the focus to *knowledge*: they could think like human experts in a specific field by possessing the same knowledge—expressed, for example, in deductive rules like the above—as the experts. To the surprise of many, the approach proved remarkably successful. One of the earliest examples, a system called MYCIN, diagnosed infectious diseases and recommended appropriate antibiotics; the system more than matched experts, outperforming members of Stanford's medical school faculty.[9] Expert systems spread throughout the 1980s, with systems appearing for recommending mortgage loans, managing toxic spill response, interpreting seismic maps for oil exploration, monitoring dam safety, and dozens of other applications. One source reports that by the end of the decade, "nearly every major U.S. corporation had its own AI group and was either using or investigating expert system technology."[10] There were also some attempts to build expert systems for education—notably, by computer scientist William Clancey—but these systems remained largely in the laboratory.[11]

Expert systems are especially valuable in situations when human experts are scarce. For example, giving one person the knowledge to carry out medical diagnostics takes many years and thousands of dollars; meanwhile, a single expert system, once programmed, can be used by any number of less-skilled healthcare workers—or even patients themselves. Moreover, expert systems can be an excellent training tool: users of expert systems can follow along as a system operates, reading the rules the system uses and thereby learning how experts reason. In other words, expert systems can make scarce, large bodies of professional knowledge more broadly available.

There is, nevertheless, a major challenge in developing expert systems: successfully extracting experts' knowledge is difficult and costly.[12] To begin with, it requires a great deal of time with the experts—time which, by definition, is scarce. A further problem arises because much of experts' knowledge is *implicit*, meaning that the experts themselves are not consciously aware of all the knowledge they have and rules they employ. Asking an expert to sit down and write down how they think will yield at best an incomplete picture, and at worst one that is outright wrong—i.e., one that does not actually reflect how the expert would reason in a real situation. For this reason, expert systems are best built by *knowledge engineers*, who specialize in extracting and organizing the knowledge of experts. A good knowledge engineer skillfully interviews experts, asking the right questions to uncover their implicit knowledge.

In the late 1990s, my father, Alex Khachatryan, thought of applying ideas from expert systems to the problem of mathematics education. Earlier in his career, he had done research on expert systems at Texas A&M University in College Station. Now, he realized that the main barrier to instructional reform—the difficulty of successfully transferring a large body of expert teaching knowledge—was *exactly* what expert systems were designed to solve. Instruction modeling was born.

The Idea of Instruction Modeling

Before undertaking instruction modeling, we must first choose an instructional tradition to model. The word "tradition" here is important: we must work with a group of teachers who all come from a single country and furthermore share a similar method.

Why is this necessary? If our ultimate goal is to transfer teaching knowledge, would it not be better to use a wide range of teachers? We could hope to benefit from the knowledge of expert Japanese teachers, Russian teachers, Singaporean teachers, and even exceptional teachers from countries where the mathematics instruction may be otherwise unremarkable. Unfortunately, this is impossible, owing to the *systemic* nature of education we discussed in the first chapter. Good instruction works as a system, with each component contributing in the context of the others. Just as in the construction of an automobile, if we pick and choose parts from different designs, we are unlikely to end up with something sensible at the end. We should model a single, coherent tradition in its entirety. It could be Japanese mathematics education, or Singaporean, or Russian, but it cannot be an amalgam.

Once a tradition has been chosen and a team of expert teachers assembled, the process can begin. The first step is to study the expert teachers to extract their knowledge. The goal here is to understand not only how the expert teachers reason but also what *experiences* students end up having in their classrooms; this makes the question a broader one, since we must seek to understand not only how students interact with the teacher, but also how they interact with textbooks, notebooks, their peers, and everything else that makes up the environment in which they learn. Most of this information can be obtained from the expert teachers—if one asks the right questions. As in classical expert systems, this work of *knowledge engineering* is challenging, owing both to the sheer volume of knowledge these experts possess and to the implicit nature of much of it. There are many different ways to go about it, each with its own strengths and weaknesses. This chapter will relate how Reasoning Mind went through several different versions of this process as it developed successive generations of programs.

Next, the instruction modeler designs online lessons that aim to recreate the experiences of students in the expert teachers' classrooms—in other words, to *model* them online. Some activities are relatively easy to model. For example, a lecture-style explanation of a mathematical concept in front of the class can be modeled as an online explanation of the same sort, presented either as text or video. But other activities are more difficult to model. For example, how do you handle a mathematical game? Or an activity in which students check each other's work? Or when the teacher calls a student up to the board to explain a problem solution to the rest of the class?

Clever interfaces can make it possible to model some of these activities. For example, Reasoning Mind's *Edifice* program for middle school uses an interface (Figure 3.2) meant to simulate a small-group lesson. In addition to a tutor character, there are two student characters. This trio has mathematical conversations, writes on the "blackboard," asks questions of the real student, and reacts to the real student's responses. This greatly expands the scope of what can be modeled. For instance, if we wish to model an activity in which a student presents a solution to the class, we can have the tutor character call on a student character to do this. The real student would watch the explanation, and perhaps be prompted to fill in steps or assess its accuracy.

At the same time, there are some important experiences that are either impossible or impractical to accurately model through an automated online lesson. For example, good mathematics education often calls for students to *explain their reasoning*, in speech and in writing; the teacher critiques the response, thereby coaching students both on how to think mathematically and on how to best express their

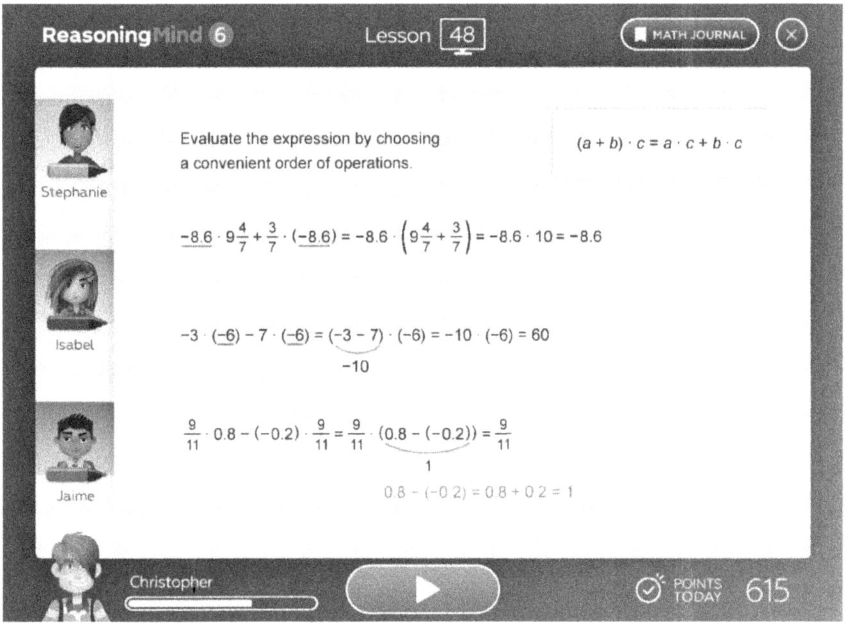

FIGURE 3.2 The interface of the Edifice program.

mathematical thoughts. To make sure this variety of experience is not lost, the final step of instruction modeling is to train and support classroom teachers so that they can deliver these experiences to students in person.

Thus, we can describe instruction modeling as a process with three steps:

Step 1. Study the experiences of students in world-class mathematics classes.
Step 2. Build online lessons that automate these experiences, replicating them as closely as possible.
Step 3. For experiences that cannot be automated, train and support classroom teachers to provide them to students.

A classroom using an instruction modeling program sees students working on computers, receiving interactive online lessons that accurately reflect the mathematics instruction of country of origin. The teacher walks around the room, working with individual students or small groups. In this classroom, students are directly exposed to the ideas, terms, explanations, problems, and illustrations of the original instructional tradition being modeled. Moreover, by reviewing the online lessons themselves, speaking with students who have done the lessons, and using program materials for one-on-one or small-group work with students, *teachers* are directly exposed to the tradition, too. In other words, this classroom has the effect of *immersing* students and teachers alike in the imported tradition of instruction. It is the closest thing to actually sending students and teachers to the country where the tradition is already in place: if successful, the effect is to provide the same quality of mathematics instruction that students would receive if we were able to magically transplant them to one of the world's best mathematics classrooms.

Instruction modeling can achieve in the realm of education exactly what expert systems are meant to do: make scarce, large bodies of professional knowledge more broadly available. Without technology, bringing expertise in mathematics education to students and teachers is almost unassailably difficult, since outstanding systems of mathematics education are rare and their associated bodies of knowledge are immense and undocumented.[13] Instruction modeling offers a solution. A program developed using instruction modeling provides a large portion of the necessary good instruction to students directly, and helps to train and support classroom teachers to provide the remainder. It is a far cry from simply giving teachers foreign textbooks and hoping for the best. Instruction modeling uses technology to provide the extensive scaffolding needed to make genuine instructional reform possible.

The Source: Russian Mathematics Education

The instructional tradition that Reasoning Mind chose to model is the Russian tradition of mathematics education. In the introduction, I told the story of my parents' disappointment with the mathematics education I was receiving in our nation's schools, and how this had led my father to the idea of instruction modeling. At the time, my parents had naturally thought that American mathematics education must be exceptionally bad, since they had nothing to compare it to besides their own experience in Russia. In fact, they were mistaken: American mathematics education is, by all accounts, average among developed countries.[14] It was Russian mathematics education—and especially the education offered in the 1970s and 1980s—that was exceptional.

The excellence of Russian mathematics education is well-known today among the world's professional mathematicians and physicists, but less so among the general public. The best source for its history is *Russian Mathematics Education*, a two-volume collection edited by historians of education Alexander Karp and Bruce Vogeli at Columbia University's Teachers College.[15] Mathematics education in Russia dates back to Peter the Great (1672–1725), the famous Russian tsar. Peter inherited a country which today we would generously describe as "developing." Russia had no mathematics or science, no expertise in engineering, and no influence on the world stage. This did not suit Peter, who styled himself the nation's first emperor and had boundless ambitions. The first ingredient Peter needed was a navy, which Russia lacked—and shipbuilding required knowledge of mathematics. Peter hired mathematicians from Germany, built schools, and commissioned mathematics textbooks.

This tradition of state support for mathematics was carried on by Peter's successors. By the end of the nineteenth century, Russia had many excellent textbooks in arithmetic, geometry, and algebra, and had produced several internationally notable mathematicians. However, the mathematics education—while good—was mostly only available to the elite, such as children of the aristocracy, military cadets, and members of the middle class with the means to pursue university study.

This changed with the Russian revolution of 1917, which brought with it a mandate to improve the education and social mobility of the working class. Universal education was enacted, and literacy spread from a small fraction of the population to virtually the entire country. Totalitarian regimes are famously erratic in their decisions, and the Soviet Union was no exception. For example, for reasons that remain murky, Joseph Stalin's government declared genetics and computer science to be "capitalist pseudo-sciences," arresting leading researchers and hamstringing work in these disciplines in Russia for generations. In mathematics, however, the regime took the opposite path: mathematics teachers and researchers were given the state's full support, and schools across the country began to offer increasingly good instruction in the subject.

Ironically, just as Russian mathematics benefited from the state's support, it also benefited from opposition to the regime. Many talented individuals became mathematicians or mathematics teachers precisely because the discipline was inherently apolitical: unlike historians, journalists, or managers, mathematicians did not need to extoll the virtues of communism at every turn, and were generally left in peace to follow their work where it led. Mathematics even held a sort of countercultural cachet: it was a discipline of pure thought, and one that was believed to train the mind to reach its own conclusions. It was held that those schooled in mathematics would be less susceptible to state propaganda. Many mathematics teachers saw their work as subtly subversive, further increasing the subject's appeal and cultural importance. Mathematics was the James Dean of the Soviet curriculum.

By the 1970s, while Russian instruction in most other subjects—such as history, biology, and foreign language—was ordinary, the country's mathematics education was remarkable.[16] Outstanding textbooks were used at every grade level, highly trained and deeply knowledgeable mathematics teachers were to be found in almost every school, and advanced students were given opportunities to attend special after-school and summer programs, participate in local and national mathematics competitions, and enroll in some of the world's best specialized mathematics schools.

This tradition was influential beyond Russia's borders. Many Baltic and Eastern European countries adopted mathematics curricula derived from the Russian ones. Russian mathematics books were translated into English and widely used—especially in programs for advanced students—in India. There is also some evidence that Chinese mathematics education drew from the Russian tradition,[17] a plausible theory given the close ties between the two countries in the 1950s. The famous Singaporean mathematics textbooks were written by authors educated in the Chinese curriculum, and their similarity in approach to Russian textbooks is substantial.

In recent years, Russian mathematics education has seen a decline. Russian mathematician Sergei Lando reports that during the last two decades, "the average level of education decreased dramatically . . . practically all experts agree that most university freshmen's mathematical background is much lower now than in the 1980s."[18]

Multiple-choice tests—once unheard of in Russian schools—have become more common. Pedagogical universities have deteriorated, unable to recruit good students. Excellent teachers have retired, with no new generation to take their place. Mathematicians have emigrated *en masse*: for example, of the seven Russian-born mathematicians to have received the Fields medal—considered the discipline's highest honor—since 1990, all but one now reside in Europe or the United States. These changes have all taken their toll. Nevertheless, remnants of the old tradition live on, at least for now: Russia still produces a disproportionate number of top mathematicians, takes top places in international school mathematics contests, and performs better than average on international tests of students' mathematics knowledge.[19] And some teachers and schools carry on the old traditions in full, using the same textbooks, lesson plans, and instructional methods as their predecessors from a generation before.[20]

First Steps in Instruction Modeling: Genie I

In the late 1990s, my father brought the idea of instruction modeling to my mother and me. What started as a quaint "family project" grew into the nonprofit organization Reasoning Mind. Our hope was to use technology to bring the best of the rich tradition of Russian mathematics education to many more students and teachers. Over the course of more than a decade, Reasoning Mind developed six blended learning programs, using instruction modeling each time (Table 3.1). Each time, some things worked well, and others badly. Here, we will tell the story of these programs: how they were made, how they were implemented in schools, what lessons were extracted, and what it all means for instruction modeling.

If there is one overarching lesson, it is that many things can only be learned through experience. Instruction modeling—and, I believe, any form of instructional design—is too complex to conduct purely in the mind. Until a program is built and tested, it is impossible to fully know its strengths and weaknesses. It will surely have unexpected examples of each, and discovering them will yield new knowledge. This is one reason why schools that use blended learning and developers who make it should never stop experimenting: it is the only path to genuine expertise.

TABLE 3.1 Reasoning Mind's programs

Program	*Grades*	*Type**	*Years in use*
Genie I	6–7	Ancillary	2003
Foundations	2–6 (grade 6 now retired)	Core or ancillary	2005–Present
Edifice	6–7 (grade 8 planned)	Core	2013–Present
Smarter Solving	2–5 (6–8 planned)	Supplementary	2014–Present
STAAR Readiness	3–8	Supplementary	2013–Present
Blueprint	Pre-K–1	Ancillary	2016–Present

*A core program is the main curriculum. A supplementary program is used at will (often sparingly) in addition to the core. An ancillary program (in the sense meant here) is somewhere in the middle: a program with its own sequence, and requiring considerable time commitment, but not rising to the level of a full core program. Reasoning Mind's core programs are typically used over a hundred hours a year, supplementary programs under thirty, and ancillary programs between thirty and seventy.

In Reasoning Mind's case, the first, tenuous step of the journey was the building of a simple prototype: *Genie I*, which drew its name from a Genie character who appeared on the screen and cheered students on. To identify expert teachers to model, we established a partnership with Moscow School #57.* Despite the unassuming name, this school is legendary, long considered Russia's leading mathematical school. Many of the country's best mathematicians number among its graduates, and only outstanding teachers are admitted to work there. The school's mathematics department chair, Boris Davidovich, and several of the school's best teachers participated in the project.

Next, we conducted structured interviews with the teachers to determine the broad outlines of how they approached teaching. On the highest level, Russian teachers divide each topic into "theory" and "problems." By *theory*, they mean explanations of mathematical ideas, analogous to the material in a textbook chapter. After

* The school has grades 1 through 11 (Russian students attend university after eleven years of school).

students have understood the ideas and looked at some examples of how they are applied, they proceed to the *problems*: independently answering mathematical questions that require an understanding of the ideas just presented. The problems can be at several levels of difficulty, with all students expected to solve the basic ones, but stronger students expected to go on to more challenging problems.

This suggested a simple structure: for each topic, students would view interactive online material giving the *theory* explanations. After this, they would proceed to easy ("Level A") problems. If students did badly on these problems, they would be flagged for attention from the teacher; if they did decently but not exceptionally well, they would move on to the next topic; and if their performance was exceptional, they would graduate to the "Level B" and "Level C" problems over the topic, moving to the next topic only afterward.

The first draft of the materials was written by the teachers, who knew that they were writing for computer lessons. The result looked something like a textbook, but with more much detail, plenty of worked examples, full explanations, and complete solutions for each problem. In other words, the materials contained all of the *content* of a lesson (which goes far beyond what is included in a typical textbook), but in a *format* closer to an interactive textbook. We translated these materials into English and edited them liberally to make them livelier and even more interactive: for example, we added animated illustrations for mathematical ideas, wrote additional examples, and broke the material up into smaller pieces that would be displayed in sequence rather than all at once.

At this stage, we were a small crew trying frantically to cobble something together. My own contribution was mostly on the side of the instructional content: I worked to translate the materials from Russian and then to edit them. Much of this work happened during the summer, while I was at Mathcamp. There were hundreds of problems—each with a detailed solution—to be translated, and my friends noticed that I took to skipping classes, avoiding Mathcamp's glamorous social scene, and generally making myself scarce on account of the workload. Several of them who knew Russian volunteered to help me out, and together we made short work of the materials. It was a touching show a friendship, and typical of how Genie I was made: improvised from the scavenged contributions of friends and volunteers.

Finally, by spring of 2003, the time came to pilot the program. Fifty students—thirty in an urban school in Houston and twenty in a suburban school in College Station, Texas—participated. Besides the theory and problems, the program included online tutors, a couple of mathematical games, and a dashboard that displayed students' study history to adults. My parents personally oversaw the implementation in Houston, acting as the classroom teachers.

It was a difficult semester, and—following the time-honored custom of beta software releases—everything that could go wrong did. Perhaps the most colorful episode took place several weeks into the program. The software, which had been surprisingly stable until then, suddenly crashed two minutes into the class. My parents frantically called the programmers, but they could find nothing wrong. After the end of class, the software returned to perfect functioning. The programmers carefully tested it, and found that everything was in order. Everyone hoped it was a mysterious, isolated event. Two minutes into the next class session, the same thing happened. My parents asked the students to leave the computers and conducted an offline lesson; the program instantly returned to perfect functioning See Figure 3.3.

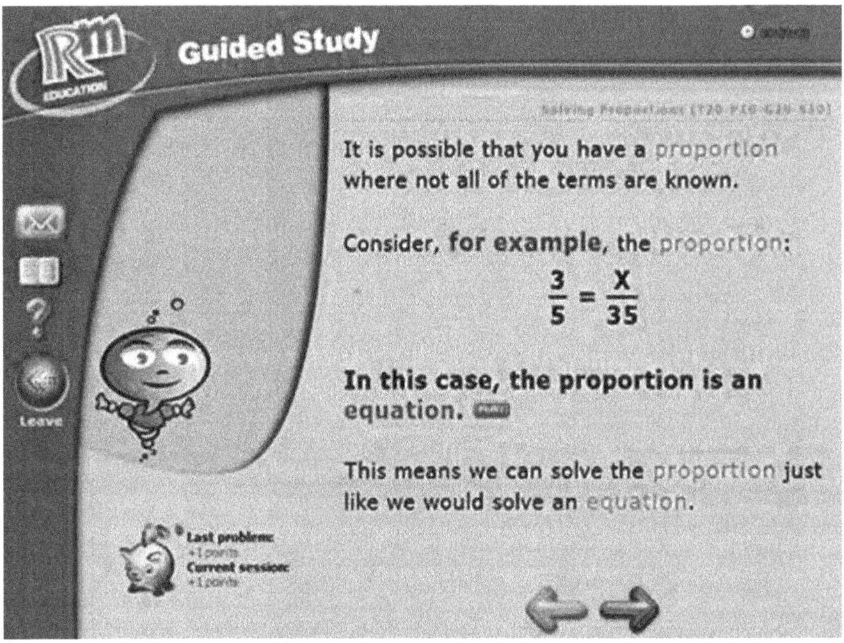

FIGURE 3.3 The Genie I program.

They asked the students to log back on, and again—the crash. My parents were at a loss. The programmers were out of ideas. But luckily, during the next class, one programmer noticed a strange phenomenon: while there were only thirty students in the class, it appeared that hundreds were logging on. It turned out that one student—no friend of mathematics and eager to disrupt the proceedings—had discovered that by clicking the "login" button repeatedly, he could crash the entire program. The program kept logging him in over and over again, creating hundreds of "sessions" in place of one. Designed to support only a few dozen sessions, the program would buckle, overwhelmed by the weight of illusory users. The bug was easily fixed in time for the next class: the button was changed to disable itself after the first press. Disaster was averted.

Despite such troubles, the implementation as a whole was a success. The program was piloted for one full semester, and by the end students had noticeably changed in their knowledge and attitude toward mathematics. An independent evaluation study was conducted by W. A. Weber, a professor at the University of Houston; the study included a randomized controlled experiment, and found large increases in student scores on three separate tests: Reasoning Mind's own test, the Stanford Achievement Test, and the Texas high-stakes TAKS test.[21]

How Instruction Modeling Benefited Genie I

In what ways did instruction modeling contribute to the success of Genie I? It had acted as a guide throughout, helping us make good design decisions. To begin with, instruction modeling had yielded a sensible way of organizing the material: conceptual explanations first, then a variety of basic problems, and finally more advanced problems to deepen understanding. Second, instruction modeling had made clear the limitations of technology: in interviewing the expert teachers, we had realized that only a portion of good instruction could be provided automatically. As one consequence of this realization, we developed a virtual classroom mode, allowing live human tutors to give students lessons on topics they had struggled with. This tutoring worked well, and clearly played a part in the program's overall success. And third, instruction modeling provided the materials themselves; the explanations were clear and mathematically sound, the topics appropriately sequenced, and the problems varied and mathematically interesting.

This last point is crucial. As the second chapter made clear, good mathematics education does not merely espouse a collection of good principles; rather, it follows these principles correctly, carefully, and consistently on every page and during every minute of instruction. Accomplishing this is far from easy. If we stopped someone on the street and gave him a list of precepts ("Pursue genuine understanding! Teach mathematical problem solving! Seek mathematical rigor!"), it is unlikely that our victim would write textbooks or deliver lessons anything like those of excellent teachers in Russia, Singapore, or Japan. This is because correctly applying the principles requires more than a superficial understanding of their meaning. It requires *mathematical maturity,* a quality one develops over many years of mathematical training. The typical first-year undergraduate mathematics major will not be able to write a rigorous proof, tell apart a mathematically interesting problem from a mundane one, or distinguish genuine from superficial understanding of a mathematical idea. By the end of her fourth year, however, she will be able to do all of these things—provided the program was a good one, of course.

But even if we asked a trained mathematician to create instructional materials following our set of precepts, we would still be unlikely to end up with something satisfactory. The textbooks and lessons of countries with outstanding mathematics education were produced over many years, battle-tested, and improved over time. Trial and error played a crucial role in the design of each and every lesson. When is an explanation too rigorous to be accessible to students of that age? When is a problem too difficult? When is an example boring? When is a discussion inadvertently confusing? These things are only learned through experience.* A course, like a software interface, must be usability tested.

The word "lesson" is important here. We often think of textbooks as containing all of the knowledge relevant to a course. But the textbook is only the tip of the instructional iceberg. A lesson lasting an hour—and whose contents would take dozens of pages to record—might correspond to only a page in a book. Much of the trial-and-error

* This is why expert mathematics teachers possess specialized knowledge *distinct* from what professional mathematicians have.

knowledge expert teachers have is not reflected in the textbook, but is nevertheless essential to excellent instruction. This is why instruction modeling must proceed lesson by lesson. It is not enough to say, "Here are ten pages describing Japanese mathematics education. Now, let's write a program that does all of this!" This approach may lead to something superficially similar to Japanese education, but it would surely lack the excellence that resides in the details. We must model *each* lesson. For every minute of a course, we must learn how *that* minute is handled, and then seek to faithfully reproduce it.

From the beginning, Genie I took this approach. The most important factor in the quality of the program was the source of its materials: they had come directly from the expert teachers, informed by their training, years of experience, and the broader tradition of which they were a part.

But despite its strengths, Genie I was only a start, the first attempt at instruction modeling. The programs Reasoning Mind developed later took the concept much further, with each successive step increasing the degree of faithfulness to the original instructional tradition being modeled.

Instruction Modeling on a Large Scale: Foundations

Following the success of Genie I, Reasoning Mind set about developing a full-scale program: one that would include enough content to occupy students for longer than a semester, and a software architecture that would support thousands of concurrent users. The resulting *Foundations** remains to this day the most widely used of Reasoning Mind's programs. The underlying approach in Foundations was the same as Genie I: using interactive animations and problems to model the practices of expert Russian mathematics teachers. However, the program saw very significant refinements both to the accuracy of modeling and to the sophistication of the instruction materials.

One lesson learned from Genie I was that teachers from elite mathematics schools were not always the best equipped to develop content for general use. Some, spoiled by their experiences working

* The program was originally named Genie II, and some journal articles describing the program refer to it by this name. I will use the current name, Foundations, throughout.

with Moscow's most motivated students, seemed to genuinely believe that no problem was too difficult to be inflicted on each and every child. Consequently, while many of the original members stayed on, new teachers were chosen not only for their expertise, but also for experience working with "ordinary" (in the teaching vernacular, "general education") students. This category of teachers proved especially hard—though, thankfully, not impossible—to find, since the best teachers in Russia are often funneled toward specialized schools serving select students. This illustrates an important principle for instruction modeling: teaching skill comes from *experience*, and therefore is dependent on the context. For example, a teacher who is expert in teaching elite ninth grade students might be awful at teaching struggling second-graders. To succeed, instruction modelers must work with teachers whose experience is as close as possible to the content and target audience of the program being developed.

With the teachers in place, the first step was to conduct detailed interviews to obtain a more complete picture of their instructional practices. This led to a much more complex instructional process for Foundations than that of its predecessor. Material that was not mastered was to be repeated. Repeated difficulty with a topic would launch a "diagnostic" sequence, which would identify—and ultimately remedy—the gaps in prior knowledge that could be responsible for the present struggle. The division between "theory" and "problems" was made less strict, as explanations and problems would often need to be intermingled. "Speed games"—activities where students would solve a collection of simple problems in rapid succession—were added to better model mental mathematics exercises carried out by the teachers.

These changes were all made to improve the fidelity of the modeling: i.e., to allow the program to better mimic the actual practices of the expert teachers. But beyond these changes, additional changes were needed to respond to specific challenges that arose during the use of Genie I. The biggest challenge was the unwillingness of students to exert any deliberate effort whenever it was not immediately required. For example, students would work to solve problems, but would mindlessly "click through" explanations of concepts. Upon solving a problem incorrectly, they would be shown a detailed, step-by-step solution—which they would ignore. They would not take notes. These oversights were not the result of laziness—although

I must confess that if the students were anything like me, surely laziness did play some part in their behavior—but rather of what education researchers call a lack of "self-regulation" skills. In other words, the students did not have the habits of experienced learners, who know that deliberate effort is needed when encountering concepts and examples, since this is the only way to acquire the knowledge necessary for solving the problems that follow.

We took these concerns to the committee of expert teachers, who confirmed that their own students were prone to the same behavior. We asked them how they handled it. They replied that the long-term goal was for students to develop self-regulation skills. However, this is neither quick nor easy: it happens over many years, arising out of the maturity that comes from years of exposure to instruction. Teachers can help it along by coaching students ("Did you take good notes? Are you sure you understand it? You'll need to understand it so you can solve the problems . . ."), but mostly it takes time and challenging content. Therefore, in the short term, compensatory measures are called for. For example, the teacher can check to make sure students are taking notes. The teacher can hold students' attention during class with the ever-present threat of spontaneous questions. The teacher can ensure students understand the corrections made to their erroneous solutions by requiring them to revise and resubmit their work.

Following the teachers' practice, we introduced "The Genie's Rules," a set of three injunctions to self-regulation: to read and understand the "theory" explanations, to read the step-by-step solution to any problem solved incorrectly, and to carefully study the notes taken during each lesson. These rules were taught to classroom teachers, displayed on wall posters (Figure 3.4), and referenced during the online lessons. But to correct the problem in the short term, stronger measures were needed. We made the "theory" explanations much more interactive, with questions asked of the student on every screen; this mimicked teachers' practice of unexpectedly calling on students during whole-group lectures. We made the detailed solutions to problems interactive, so that students who solved a problem incorrectly would need to go through each step of its solution, filling in blanks to demonstrate understanding. We added a "notes test" to each lesson, emulating a notes check that a teacher might conduct. Collectively, these changes took an enormous

FIGURE 3.4 A poster showing The Genie's Rules. Printed copies of this poster were provided to all Reasoning Mind teachers.

amount of work, since developing highly interactive content is far more laborious than making analogous static content. However, they succeeded in solving the problem: Foundations students can no longer mindlessly "click through" material with impunity, nor can they complete a course without a dose of coaching to support the development of self-regulation skills.

The matter of self-regulation illustrates an important aspect of instruction modeling: far from being best in isolation, instruction modeling *benefits* from combination with other approaches. In this case, the self-regulation deficiency was first identified through *iterative design*, a cornerstone of modern software development; this practice calls for building any system in stages, testing each stage and improving it based on what the testing shows. Once the deficiency was identified using *iterative design*, we turned to *instruction modeling* to find an appropriate solution. We ended up with a solution that improved the fidelity of modeling: i.e., was a better mimic of the expert teachers' practices. But this would not have happened had we not relied on iterative design in the first place, since we would likely have overlooked the importance of the problem. The two methods worked hand in hand, to the benefit of each.

The Limitations of Modeling in Foundations, and the Response: Edifice

When Foundations was released in 2005, it included only a course for fifth grade mathematics. By 2011, courses had been added for the second, third, fourth, and sixth grades. Grades 2–5 were successful, but the sixth grade course was worse-received: students complained that the program's appearance was too childish for them, and—what is worse—struggled with the course materials. One problem was that as the mathematics had become more involved, screens had become overburdened with text and consequently overwhelming for students.

My colleagues and I at Reasoning Mind decided that the logical next step was to develop a complete middle school course, using a new platform that we would design from scratch. This program would be age-appropriate to middle school students. More importantly, this was an opportunity to take a new direction, learning from our prior experiences. The resulting program, named Edifice (to tie in with the name of the elementary school program, Foundations), was Reasoning Mind's largest departure from prior practice since at least the release of Foundations.

Two main angles were taken in designing the Edifice program. The first angle was to more systematically apply learning science to the design of the programs; this is discussed in detail in chapter 4. The second angle was to improve the *fidelity of modeling* over Foundations. In other words, we knew that Foundations suffered from some shortcomings in its ability to accurately recreate the original offline lessons, and we wanted Edifice to do better. Put another way, we wanted to improve the quality of the *knowledge engineering*.

We had noticed that many important things were being lost in translation when we asked the expert teachers to write materials directly for the offline lessons. Some aspects of their practice were being lost out of simple inattention. For example, a teacher might give helpful tips and commentary during a live class: "Make sure you take careful notes in this part, since it is especially important." When writing an online lesson, the same teacher might neglect to write in this comment, since the frame of mind of writing an online lesson would push the teacher away from the informal—but nevertheless essential—aspects of their work, leaving behind something more akin to a textbook.

Another source of inaccuracies arose from cases when the expert teacher did not straightaway see a way to model a certain activity online, and thus would discard it or substantially modify it. For example, in a live class the teacher might ask students to check one another's work. This can be modeled in an online lesson by introducing a character ("Meet Higgie the Giant Hamster! Now, please check her answers to these problems."). But this might not occur to the teacher writing the lesson, and instead he may simply convert the activity into a set of exercises, thus substantially altering the activity's role in the lesson.

A final source of inaccuracies was forced by the constraints of the program itself. For example, in a live class a teacher might show an example problem and then discuss in great detail how he would go about solving it. Attempting to do the same online in Foundations, which has no provision for sound besides the reading aloud of text from the screen, would yield a sequence of screens brimming with text—something that would at best be ignored by students, and at worst would cause frustration and helpless resignation. Accordingly, the only option would be to present a substantially condensed version of the problem solution, including only the key steps and omitting all of the detailed commentary and explanations. This variety of inaccuracy—i.e., the kind forced by the limitations of the program—is always inevitable to some extent. There are, after all, those essentially human activities that cannot be automated under any circumstances. Nevertheless, a good instruction modeler should always seek to reduce the constraints imposed by technology, aiming to build a program that can accurately model as wide a range of relevant instructional experiences as possible.

In all of these cases of modeling inaccuracies, we did not find out about them until it was too late. After all, the expert teachers were solely responsible for translating their experience to the online lessons, and any compromises, omissions, or modifications took place within their minds, recorded nowhere and shared with no one. True, there was a process of extensive editing after the teachers finished their work, and the editors would often speak with the teachers to inquire about the materials. Some distortions would be caught and rectified at this stage. However, this was the exception: on the whole, the task of accurate modeling lay almost exclusively with the expert teachers themselves.

This went against the practices recommended by computer scientists involved in knowledge engineering in other fields. After all, subject matter experts were experts in the subject matter, *not* in the accurate modeling of their work on computers. Accordingly, the recommended approach was to have a separate *knowledge engineer*, whose job was to interview the subject matter expert and then to design the computer system; the subject matter expert should, in fact, be kept at arm's length from thoughts about computers, lest she distort the accuracy of her descriptions of how she reasons.[22] The knowledge engineer would be best equipped to extract the expert's knowledge, including all of the *implicit* knowledge the expert holds—knowledge that the expert takes for granted, and thus would never think to share unless explicitly asked. We knew all of this, but had never had time to set up such a complex approach. Moreover, we had assumed that the distortions were relatively small. After all, it is natural to think that the act of writing materials for a computer lesson should be very similar to the act of teaching a live class.

By the time we were beginning to plan Edifice, we had realized the error of our ways. This came about over several years, as we came to trace specific gaps in students' skills back to inaccuracies in the modeling. For example, we had noticed that fewer students were successfully solving advanced problems (the "Level B" and "Level C" problems mentioned earlier) than we had expected. We asked the expert Russian teachers: "Do you agree that not enough students are learning to solve these problems?" They agreed. We asked, "Do students in your classes learn to solve them better?" They did. We asked, "Does this surprise you?" It did not. We asked, "Why not?" And then the bombshell: "There are all kinds of things we do in our classes to help students develop that skill. But we don't do those things in the online lessons, because we don't see any way we could. So why would we be surprised?"

One particularly salient example is an activity called *work at the board*. This activity, which takes place during almost every Russian mathematics lesson, is seen by Russian teachers as especially important for developing students' ability to solve complex problems. In work at the board, the teacher calls a student up to present his solution to a problem. The student writes the steps on the board and then explains each one. The teacher questions the student: "Why did you do it this way?" "What rule are you using here?" "What is the next step?" The rest of the class observes the exchange, participating by offering suggestions or pointing out mistakes.

Work at the board is two activities in one, since the student at the board and the students watching from their desks have distinct experiences, each beneficial in its own way. The students watching see an example of how to reason. After all, the discussion between the teacher and the student at the board is not a finished solution, but rather a living process of working through a problem. It is presented with all of the mistakes, hunches, dead ends, backtracks, and surprises inherent in solving complex problems. By watching the conversation, students learn to repeat that same conversation inside their own minds as they work on new problems. In other words, they are shown *ways of thinking*, not ready-made answers. Meanwhile, the student *at* the board learns a different set of skills: how to present his thoughts clearly, how to defend his ideas, how to structure a written solution. Perhaps most importantly, even though a given student might only be called to the board once every two or three weeks, every student knows that such a possibility always exists. Thus, he looks differently at every problem encountered in the class, thinking not only of how to get the correct answer but also of what he would think and say when the fateful moment arrived. In other words, the possibility of being called to the board expands every student's thinking from the purely procedural to the conceptual, from *doing* to also *understanding*.

The experience of the student at the board is a classic example of an experience that cannot be automated. After all, the teacher must understand the student's reasoning, respond to it with meaningful criticism, evaluate the clarity of the student's written and spoken ideas, and carry on an undeniably complex conversation. The experience of students watching from their desks, however, can in principle be modeled in an online lesson: they are mostly passive observers, called on to furnish some missing steps or to evaluate the correctness of specific statements, but not required to participate fully in the conversation. Foundations, while it offered much, did not offer the flexibility needed to model this experience of observing work at the board.

Work at the board affords just one example. We had learned about many important activities that were too complex to be modeled with the tools we had already developed, and we knew that there must have been many more such gaps that we had not yet found. Moreover, we had come to realize that asking expert teachers to be their own knowledge engineers was introducing more substantial distortions than we had thought, and—what was worse—was doing so in a manner that hid the distortions from our view. Edifice offered an opportunity to advance our practice of knowledge engineering, substantially increasing the fidelity of modeling. To this end, we decided to redesign our process for developing online lessons, establishing a role of knowledge engineer separate from the expert teachers. We also decided to build the Edifice student interface in a way that would dramatically expand the range of experiences the program could accurately model.

The process we designed (Figure 3.5) begins with the team of expert teachers. The teachers agree on the broad outlines of the whole course: what material will be covered, what sections it will be divided

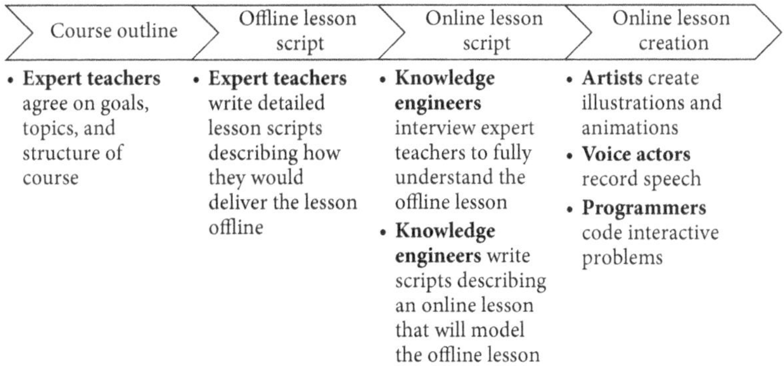

FIGURE 3.5 The lesson development process used in Edifice.

into, what skills will be developed. Each teacher is then assigned a set of lessons, and proceeds to write detailed descriptions of how he would deliver each one; these are often over fifty pages each, and describe everything in great detail: what the teacher says, what he writes on the board, what activities he has students do, how he responds to various student mistakes or questions, and so on. These descriptions never reference computers—they describe the original, offline lesson that the teacher would deliver, not how it should be modeled online. For a course of 120 lessons, this amounts to roughly six thousand pages of information.

Next, each lesson is assigned to a *knowledge engineer*. Reasoning Mind's team of knowledge engineers, which was created when the project to develop Edifice was begun, consists mostly of mathematicians, always with bachelor's degrees and usually with doctorates in the subject. When assigned a lesson, a knowledge engineer begins by reading the offline lesson description written by the teacher, and then schedules a phone conversation with the teacher to ask questions. She might ask: "If the students were better prepared, would you still explain this the same way? How would you respond here if the student gave a wrong answer? Why did you use this activity here, instead of something else?" A Russian-English translator joins the call. The knowledge engineer seeks to extract all of the professional knowledge the expert teacher employed in designing the lesson. The team's unofficial motto states their ambitious goal: "To understand the lesson on the level the teacher understands it."

When the knowledge engineer feels she has met this charge, she proceeds to design the online lesson. This is where the new student interface comes in: Edifice offers great flexibility in faithfully modeling a wide range of activities. While Foundations displays problems and explanations through text and images, somewhat akin to a highly interactive online textbook, Edifice instead has an interface (Figure 3.6) meant to simulate a small-group lesson. In addition to a tutor character, there are two student characters, each represented by an icon on the side of the screen. Each character can speak, write on the "blackboard," and use a pencil to gesture to content on the board. For example, a character can write "$x + 5 = 7$" on the board and then say, "Let's solve for x," pointing to "x" on the board with his pencil as he speaks. Thus, in Edifice sound and text are no longer yoked together: some things are written and other things are spoken, just as

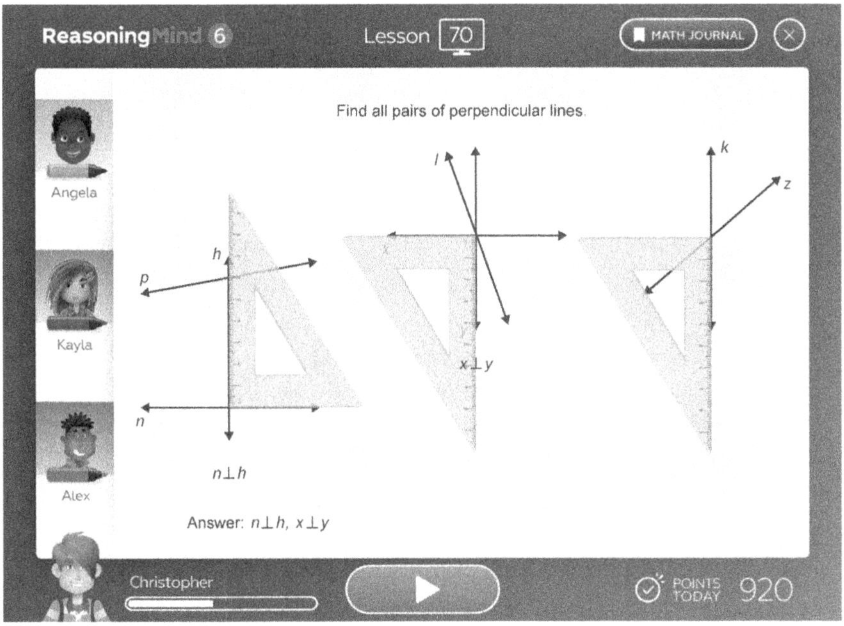

FIGURE 3.6 The student interface in the Edifice program.

in a live classroom. Just like Foundations, Edifice is not a one-way channel: the lesson pauses frequently (on average, once every thirty seconds) and the student is asked a question; the student enters the answer and is given feedback. The three characters can ask questions of the real student, react to the real student's responses, and carry on mathematical conversations among themselves.

The particular things that the characters write and say are scripted individually for each lesson by the knowledge engineer. The questions asked of the real student are also scripted, as are the possible responses to different answers. For example, the knowledge engineer might have the tutor character ask, "What is the sum of 13 and 3?" Then an answer blank would appear on the screen and the student would be prompted to type in the answer. If the student enters "16," the tutor character praises her and moves on. If the student enters "10," then a student character might intervene to call out the mistake: "I think you're subtracting there! The problem was to add." Then the real student would have a second chance to answer the question. Of course, this is only an example: the knowledge engineer has complete flexibility to include any or none of these interactions.

This flexibility is certainly a great strength, but it also means that developing each lesson is a large undertaking. At the time of this writing, each knowledge engineer worked on a lesson for three weeks, full time. The scripts that the knowledge engineer writes—scripts that dictate the online lesson's behavior and read something like an interactive play—make up an average of fifty pages. This is in addition to the fifty pages written by the expert teacher. Thus, a hundred pages have been written at this point. For a course of 120 lessons, this adds up to twelve thousand pages. If these pages were printed and placed in binders, they would occupy not a shelf, but an entire bookcase. And this documentation is only what is produced *before* each lesson is sent to the team which actually builds it. That process is even more complex: artists illustrate each screen, voice actors record the characters' speeches, and programmers code the lesson's intricate flow, complete with all of the interactive questions and possible program responses. All told, a team of several dozen people works intensely for a year to produce a single course.

Edifice's "small-group lesson" interface greatly expands the scope of what can be modeled. For instance, if the knowledge engineer wishes to model the experience of watching work at the board, she can have the tutor character call a student character to the proverbial board, asking the character to solve a problem. The student character would write on the virtual board, explain his reasoning, and respond to questions from the tutor character. The real student would watch the discussion, prompted along the way to fill in steps or assess the accuracy of statements made by the student character.

Many other activities can be readily modeled with this interface: having students check each other's work, mathematical games in which students compete to see who arrives at the answer to a problem first, mathematical disagreements that must be adjudicated by students, and much else besides. Of course, the real student can only communicate through simple inputs, such as typing in numbers, selecting from a list of options, or dragging and dropping elements on the screen. But besides this limitation, the interface affords an experience that is remarkably close to being in a live class.

Chapter 4

The Role of Learning Science in Instruction Modeling

INSTRUCTION MODELING IS NOT MEANT TO BE USED IN isolation, but rather has natural symbioses with other design practices. For example, Frederick Hess and Bror Saxberg have advocated for *learning engineering*, by which they mean the systematic application of learning science* to the development of instructional programs. A learning engineer makes it her business to know the design principles uncovered by laboratory research in education, and ensures that the programs she designs conform with these principles. Learning engineering and instruction modeling make for a pairing that even the most fastidious sommelier would admire. Each can guide us on topics where the other is silent. In cases when both offer an opinion, apparent disagreements are often symptoms of our own misunderstanding of one or the other, and help us to identify and correct these misunderstandings.

* "Learning science" can be defined as the discipline that applies the scientific method (including the use of experiments and the quantification of outcomes) to the creation of knowledge about learning.

Instruction Modeling. George A. Khachatryan, Oxford University Press (2020)

DOI: 10.1093/oso/9780190910709.001.0001

How Learning Science Complements Instruction Modeling

The first release of Foundations had not been designed with learning science in mind. However, it nevertheless ended up conforming to a wide range of scientifically derived principles for good instruction. Skills were emphasized, but practiced in small installments rather than in large blocks. Conceptual understanding was sought before the corresponding procedural skills were developed. A wide variety of challenging problems were presented in order to support the development of general problem-solving skills.

These are only a few examples, and many more could be offered. What explains this seeming coincidence? In fact, it is no coincidence. True, instruction modeling draws not on the quantitative tradition, but rather on the humanistic study of excellent mathematics teaching. Nevertheless—as we saw in the first two chapters—on questions where humanistic and experimental approaches both offer an opinion, the two are generally in agreement. Thus it should be no surprise that a program that is based on the practices of outstanding teachers should also conform with the recommendations of learning science.

Since its first release in 2005, Foundations underwent substantial yearly expansions and improvements. Concurrently, as the design team learned more and more about learning science, we found opportunities to apply it in cases when instruction modeling was of no help. One example concerns the "Next" button, a seemingly innocuous button that has been the cause of great controversy within Reasoning Mind's walls. This button is used by students to move forward through the material. In the first release of Foundations, the location of the button varied: sometimes it appeared immediately below the instructional content area, while other times it moved around to different locations within the content area itself. There were even times when there were *two* different "Next" buttons—an inactive one below the content area, and an active one on it (see Figure 4.1).

We had some evidence that this confused students: they had no way of predicting where the "Next" button would appear, and thus would need to search the screen for the correct button every time the content paused. The design team was conflicted on what

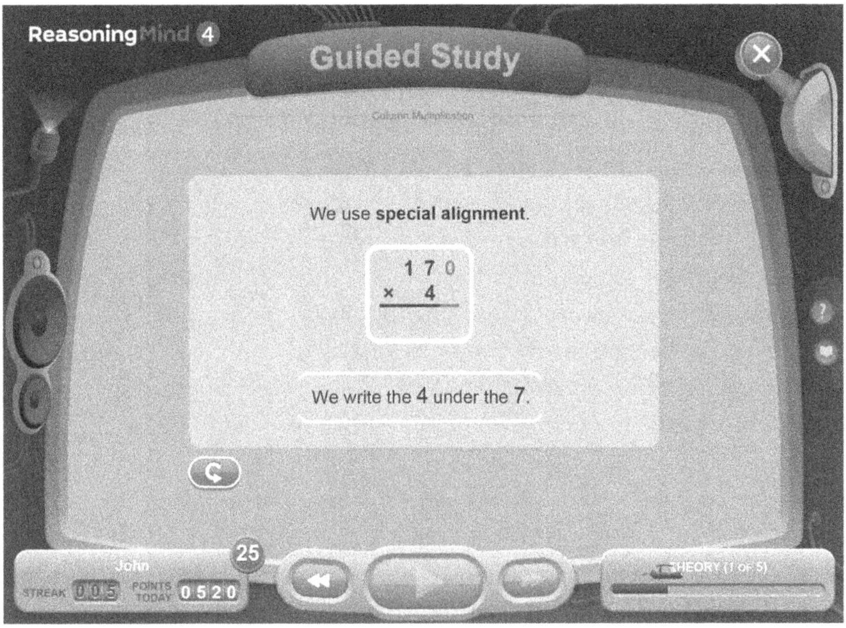

FIGURE 4.1 The original version of the Foundations interface, with two different "Next" buttons (indicated with black arrows) appearing in different parts of the screen.

to do. Fixing the problem would require a massive effort, since it would entail reprogramming each instructional animation. Some team members believed that this was worth the effort; others believed that it would be an improvement, but not enough to merit the investment; and still others claimed that the current way was instructionally better, since the need to choose between the two buttons would slow students down and give them more time to think about the material. Teacher interviews could not resolve the controversy: classroom teachers are equipped with no "Next" buttons at all, and could offer nothing beyond subjective opinions—which were already in plentiful supply.

Fortunately, learning science offered guidance, ultimately tipping the balance. Psychologists had for many years studied *working memory*, which is the mental space we have available for active thought. For example, adding 8 and 7 mentally requires holding both numbers in the mind, breaking 7 into 2 and 5 (while continuing to hold 8), reassembling to combine 8 and 2 into 10 (while still holding 5), and finally combining 10 and 5 to obtain 15. Doing any calculation or making sense of any new information requires the use of working memory. Sadly, we have but little of it: psychologists estimate that we can hold only about seven units (whether numbers, words, or other items) in our working memory at one time. If you have ever sat in a lecture and felt overwhelmed by the material, unable to keep all of the necessary pieces in mind at the same time, then you have experienced the consequences of our limited working memory. Psychologists consider this a major "cognitive bottleneck," one of the largest limitations on our ability to reason and learn. Because working memory is so precious, anything that occupies it without profit is detrimental to learning. For example, background music while learning may seem harmless enough, but has been shown in many studies to take up working memory students cannot afford to forfeit, diminishing their ability to learn. In the case of the "Next" button, the implication was the same: searching for the "Next" button occupied working memory for a task extraneous to the substance of the lesson. It was surely harmful. The animations were all redone, and a "Next" button has never since dared to appear anywhere besides the bottom-center location on the Foundations screen.

This example illustrates an important general distinction. There are two types of decisions that must be made in developing online learning programs: decisions concerning the *content,* and decisions concerning the *interface*. By *content* I mean everything that is independent of whether the material is delivered in person or online, through text or through speech, with gestures or without, and so on. For example, the choices of which problems to assign, what specific explanations to offer, and what response to give to incorrect answers are all examples of *content*. Meanwhile, the *interface* includes the particular ways in which this content is delivered, independent of its actual substance. For example, the question of whether to consolidate the "Next" buttons into a single button is a question of *interface,* since the substance of the instruction would be identical in either case.

Teacher interviews can serve as a comprehensive guide to questions of *content,* but provide no reliable guidance on questions of *interface*—after all, in the course of moving from traditional instruction to online lessons, the interface is inevitably altered to something with which the teachers have no experience. Learning science serves to fill this void. It is a stark example of the complementary relationship between experimentally and humanistically derived knowledge in education.

Edifice: Applying Mayer's Cognitive Theory of Multimedia Learning

Learning science had been valuable in making improvements to Foundations, but had not played a central role in the program's initial design. The Edifice program, which was designed years after Foundations had been first created and implemented in schools, offered an opportunity to incorporate the recommendations of learning science from the very beginning. Improving the fidelity of modeling had been the one major goal for Edifice; improving the application of learning science was the other.

In the previous section, I introduced the distinction between the *content* and the *interface*, the former being the substance of the instruction and the latter the specific form in which it was delivered. For example, the same discussion given as text versus voice offers identical content but has a different interface. Teacher interviews serve as a source for *content*, but give little insight as to the best *interface* to use. We turned to learning science.

The most important tool for answering the questions we had was a theory called the cognitive theory of multimedia learning (CTML).[1] This theory is famous among education psychologists but less widely known among designers of educational programs. While many researchers have contributed to the CTML over the years, its central architect has been Richard E. Mayer of the University of California, Santa Barbara. Mayer is the author of five hundred publications, including thirty books, and was ranked the most productive educational psychologist in the world by the journal *Contemporary Educational Psychology*. Those of us who design instructional technologies are fortunate that Mayer has chosen this topic as the focus of his research.

The CTML takes as its starting point a model of the mind—including, for example, the limited nature of working memory discussed in the previous chapter—and draws from this model a series of consequent design principles. Each principle has been tested in many laboratory studies by Mayer and his colleagues. While the extent of the evidence varies across the principles, all of them are generally well supported.

I will not list all of the principles here, since not all of them were relevant for the decisions we needed to make. For example, the *multimedia principle*[2] states that graphics and words should both be used, as words alone fare worse in stimulating students' thinking; since graphics were already extensively used in Foundations and were essential for modeling the work of the expert teachers, this was never at issue. Many of the other principles, however, provided valuable guidance in situations where the correct course of action was unclear (Table 4.1).

The CTML's *modality principle*[3] states that words should be presented as sound (rather than text) in any situation where they are referring to graphics; this is because learners process sound and visuals separately, so attempting to send everything through just one channel is overwhelming. In a mathematics lesson, the "graphics" are not only geometric diagrams: they can be tables of information, calculations, or equations. In other words, almost every minute of a typical mathematics lesson consists of verbal commentary on visual "graphics," as formulas are written on the board but then discussed through speech. The modality principle is especially important in cases when the material is complex. In a sense, the principle confirms what we already had observed: that as the sixth-grade material in Foundations had become more complex, students became overwhelmed with the text-filled screens. The modality principle offered scientific confirmation of our hope that adding sound could fix the problem. Of course, this would require a considerable departure for the approach of Foundations, which did not include sound except for narration of on-screen text. But it was clear that this effort was necessary.

A third principle is called the *personalization principle*.[4] The name is unfortunate, since the word "personalization" is so overburdened with different meanings that Jeremy Roschelle and Nikki Shechtman, two researchers at SRI International, have kept up a collection of its various uses. So reader beware: when someone speaks of "personalization" in the context of blended learning, it may have nothing at all to do with this particular sense of the word. In Mayer's meaning, "personalization" means that material is delivered in an informal

TABLE 4.1 Selected Principles from the Cognitive Theory of Multimedia Learning, and their Implications for the Design of Edifice

Principle	*Description*	*Implications for Edifice*
Multimedia principle	Graphics and words should both be used	Graphics (e.g., mathematical diagrams) are used throughout
Modality principle	Words should be presented as sound (rather than text) in any situation where they are referring to graphics	Explanations are presented verbally by characters in the program (instead of everything being presented through text)
Personalization principle	Material should be delivered in an informal style and a friendly voice, creating the illusion that it is being conveyed through a social interaction	All material is presented by convincing characters, who address the learner directly
Contiguity principle	Graphics and words (whether printed or spoken) should be coordinated	Gesturing with the "pencil" is used to indicate those parts of diagrams or formulas that are being mentioned verbally at each moment

style and a friendly voice, creating the illusion that it is being conveyed through a social interaction.*

This last point is subtle, and bears further explanation. Our minds perceive some experiences as social interactions. When someone greets us in the hallway, or we speak over the phone with a friend, or a teacher explains a concept to us, our minds flip a proverbial "social

* Perhaps the *mot juste* for this principle would be *personification*. But the ship appears to have sailed.

interaction mode" switch. By contrast, when we open a can of pickles, or close the garage, or hear a building fire alarm, our minds perceive us to be engaged in a nonsocial interaction—the sort we ordinarily have with the inanimate world. But modern technologies have blurred the line, sometimes flipping the social interaction switch when no actual living beings are involved.

My favorite example of this concerns a particularly advanced trash container at a Cornell University dining hall. I would often eat lunch there, and at the end of my meal would dispose of the leftover paper plates. As I approached, the trash container would automatically swing open, receive the trash, and thank me in a recorded human voice. Each time, I felt a desire to reply: "Of course, don't mention it. Thank *you*, really." At the time, I thought this was evidence of particular whimsy on my part. Now, I know the truth: everyone in my position would have felt the urge to respond. The solicitous container had done just enough things to flip my social interaction switch: exhibited motion in opening, spoken in a friendly human voice, followed social conventions of politeness. If it has neglected to do one of those things—remained motionless, or used a badly synthesized voice, or said "trash deposited" instead of "thank you"—it would likely have been insufficient, and I would have been left unmoved. But as it stood, the trash container had succeeded. I knew perfectly well that this was not a social interaction, that we would not confide in one another or become fast friends, but my mind did not care: it had decided, and that was that.

This is relevant because, per the CTML's personalization principle, *we learn better* when this social interaction switch has been flipped. For some mysterious reason, if we take an online lesson written in an impersonal style and make changes to make it feel like a social interaction—rewrite it in the first and second voice, add informal or polite speech, give the speaker's voice some personality—then students actually learn more from the lesson. In other words, if that Cornell trash container had followed up with a lecture on the history of waste disposal, I would likely have understood and remembered it better than if I had encountered the exact same material in a Wikipedia article.

For Edifice, the personalization principle meant that material would need to be presented by socially convincing characters. Moreover, they would need to speak in real voices, not synthesized

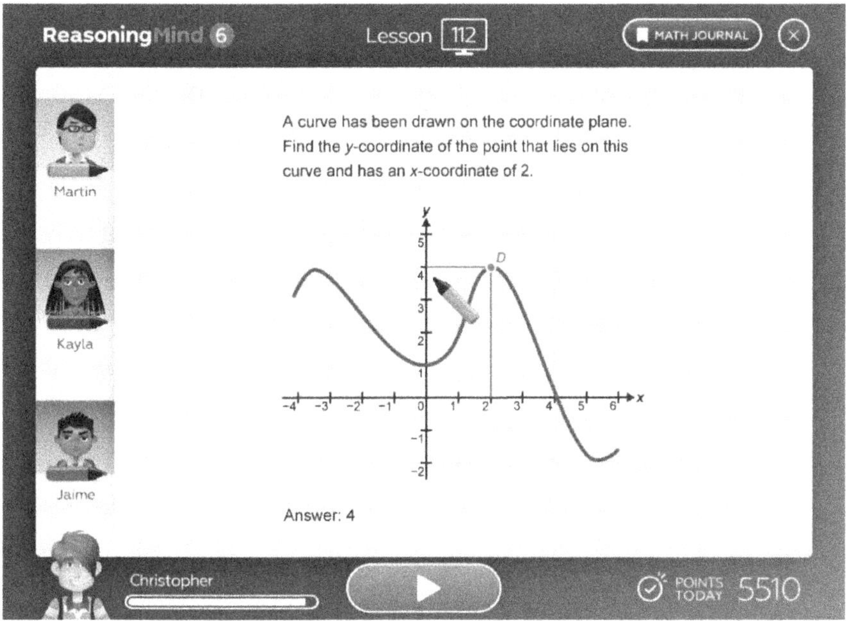

FIGURE 4.2 The Edifice interface, which aims to systematically apply the recommendations of the cognitive theory of multimedia learning.

ones, since the research suggested that synthesized voices erased the personalization effect. Accordingly, we designed the interface to simulate a small-group webinar, with icons for a "tutor" character and two "peer student" characters* on the left, and a virtual blackboard on the right (Figure 4.2). The real student could design an avatar, which was displayed on the bottom. Following the modality principle, the characters would all speak to one another and to the student, verbally discussing graphics (including formulas, diagrams, and important facts) shown on the board.

A natural question we asked was whether the icons should also be animated—i.e., whether the lesson should simulate a group videoconference. The answer was suggested by the *contiguity principle*,[5] which states broadly that graphics and words (whether printed or spoken) should be coordinated. For example, presenting a graphic first and a verbal explanation after the graphic is no longer displayed

* These peer characters were needed for improved fidelity of modeling, not because of anything in the CTML.

would violate this principle. Another example of a violation of the principle would be displaying a graphic on one side of a screen and a printed commentary on the other, leaving it to the student to link the two. The reason for this principle is that splitting associated graphics and explanations forces students to store one or the other in working memory, thus crowding out space which they need for making sense of the content. Animated icons would violate this principle in a subtle but important way: the motion draws the student's eyes away from the material on the virtual board, thus breaking the coordination of words and visuals. Ignoring this is a common design error in many online learning programs, which often display a slide presentation along with a video recording of the speaker in the corner.

There is another important implication of this principle. Lecturers use gesturing to indicate relevant aspects of graphics as they speak. For example, a teacher might point to "*x*" on the board as she says, "We must solve for *x*." In an online lesson, it is easy to omit this form of gesturing. However, omitting it forces students to mentally search for the visual items being referenced by the speaker, taking up working memory in the process. Thus, some form of *gesturing* helps to better link associated words and images, and thus serves the contiguity principle. Following this, Edifice equipped all of the characters with pencils, which they could use both to write on the board and to gesture to items as they referred to them (Figure 4.3).

It did not occur to us at the time, but Richard Mayer later pointed out to me that the pencils end up serving a second purpose: their motion helps reinforce the personalization effect. In other words, because the pencils' motion is associated with the characters, it makes the mind more likely to perceive them as actual social beings.

Beyond the modality, personalization, and contiguity principles, there were of course many other recommendations from learning science—both from the CTML and from other corners of education psychology—which informed the design of Edifice. But these three were the most consequential, and also serve well to illustrate the specific way in which learning science can be applied to program design.

There is an interesting perspective on the CTML beyond how it is usually presented in books and scientific articles. When we hear the word "multimedia," we think immediately of technology. But traditional, offline instruction—even something as one-directional as a lecture—is actually a very rich multimedia experience. The

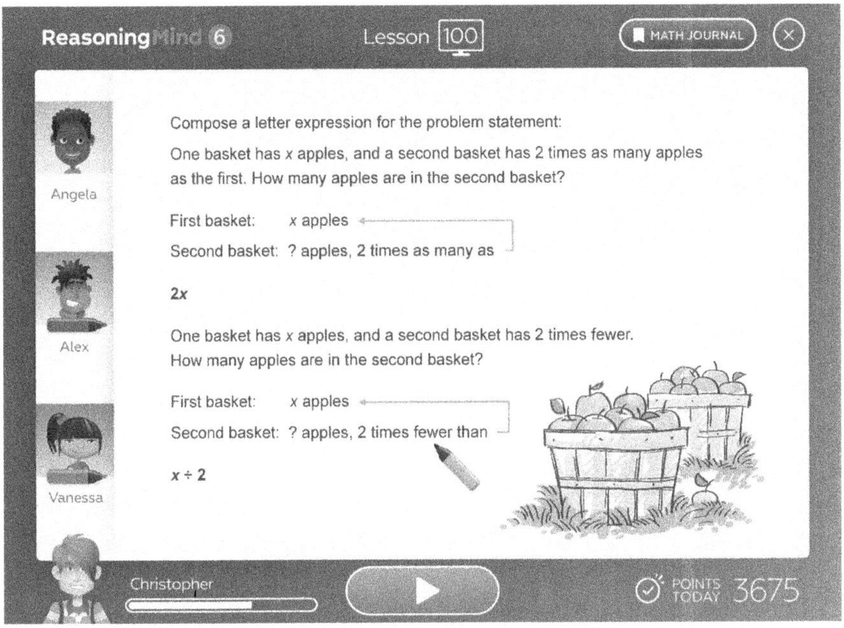

FIGURE 4.3 Edifice characters use their pencils to point to elements of the screen as they refer to them in speech. This gesturing conforms with the *contiguity principle.*

teacher speaks some things, and writes others on the board; charts and diagrams are drawn; the teacher gestures to emphasize his points or to draw students' attention to something that was displayed; the experience is intensely social, with the teacher telling stories and relating to students; the teacher asks and answers questions; the teacher's and other students' faces are visible; the teacher has a presence in three dimensions. Which of these things are to be found in the typical online learning program? Usually, very few. Including all of them would be technically impossible, and at any rate prohibitively expensive, as we would need to program fully animated, interactive lessons in virtual reality.

Therefore, we must accept that our online learning will be *less* rich as multimedia than traditional instruction. This leads naturally to the question: which multimedia aspects of traditional instruction matter for learning? For example, if we find that the three dimensionality of the teacher is not really so important, then we can safely forget about using virtual reality—and save ourselves a great deal of work and expense. The CTML answers precisely this question. All told, it suggests that online learning needs to retain most—though not all—of the multimedia features of in-person instruction if it is to be as good. We do not need to go all the way to virtual reality with convincing, three-dimensional teacher characters. But we also cannot content ourselves with interactive textbooks.

This interpretation of the CTML is especially appealing to an instruction modeler. After all, our aim is to study outstanding offline instruction and then use blended learning to bring equivalent educational experiences to students; teacher interviews help us provide equivalent *content,* but do not help us find the right *interfaces* to ensure that the experiences are truly equivalent. The CTML steps in to fill the gap. From this perspective, the CTML is not something that is applied in *addition* to instruction modeling, but rather something that is applied as *an essential part* of the process of instruction modeling.

All of this illustrates the relationship between experimental and humanistic approaches in blended learning, and in education more broadly. Far from being in opposition, they complement and support one another. Understanding both is crucial for designing the most effective instruction for students.

Chapter 5

Implementation
When the Rubber Hits the Road

THERE IS A LESSON THAT FEW EDUCATIONAL PROGRAM designers seem to learn in any way besides the hard way: the greatest challenge in any instructional program is the challenge of *implementation*. It is one thing to create a program that functions well in a laboratory or a single classroom, and another thing entirely to expand such a program to a large number of schools without losing what had made it effective in the first place. No program can succeed without a thoughtful approach to meeting this challenge.

But as much as it is a challenge to be overcome, implementation is also a remarkable source of knowledge. In this chapter, we will discuss both the difficulties and the lessons of implementation, with a special focus on the implications for programs designed using instruction modeling. Because instruction modeling programs are by their nature comprehensive, the challenge of implementation is in their case especially acute.

Instruction Modeling. George A. Khachatryan, Oxford University Press (2020)

DOI: 10.1093/oso/9780190910709.001.0001

The Crisis of Implementation and the Essential Role of the Classroom Teacher

In 2003, after the extraordinarily successful piloting of Reasoning Mind's first system Genie I, the Reasoning Mind team was convinced that we had solved the problem of providing excellent mathematics instruction at a large scale. All that remained, we reasoned, was to build a version of the program that could support more users. We did not realize the many advantages the Genie I project had. There were, of course, the intrinsic qualities of the software: it was clean and simple, it had a sufficient variety of activities to interest students, and it had a strong underlying curriculum. But the distinguishing factor was the strength of the *implementation*: having teachers in the classroom—in this case, my parents—who knew the material deeply had a remarkable effect. We did not realize it at the time, but these were the best possible conditions for implementing a new approach.

When Foundations—our first program intended for broad use—was released to three schools in the fall of 2005, we were headed for a rude awakening. We provided basic software training to teachers using the program: just enough for them to know where the buttons were, but no discussion of the underlying curriculum, how to best implement the program in the classroom, or anything else. Throughout the year, we occasionally visited classrooms, but there was no structured process to ensure that the implementations were going well. We knew that the implementation quality was low, but somehow still expected the results to be good.

At the end of the year, W. A. Weber, the same independent evaluator who had found striking results in the Genie I project of 2003, analyzed the data from the 2005–2006 school year and found no evidence of impact. Only one test (the Texas TAKS) had been used, so for the reasons discussed in the first chapter, this alone was far from enough to draw any definite conclusions one way or the other. But the other evidence seemed to point in the same direction: teachers were lukewarm about the program, the time students spent actually using the program was low (averaging only several dozen hours per student), and students had only completed a fraction of the course material. Fortunately for Reasoning Mind, Weber knew more about educational programs than we did: he observed the implementation flaws, and noted that they alone were enough to spoil an otherwise effective program. On the force of his recommendations, donors and schools agreed to a second attempt, this time with a focus on implementation quality.

Incidentally, this crisis beautifully illustrates the value of field experiments in education. Even with all of their limitations, the blurry snapshots they provide can be a powerful inducement to constructive action. If a well-designed study had not been conducted, we might very well have gone on thinking that the implementation flaws were of minor importance. In the absence of concrete countervailing evidence, people prefer to believe what is more soothing. With Weber's study suggesting a serious problem, we were forced to confront the deficiency—to the ultimate benefit of the program and the students who used it.

From that point on, implementation became a central concern at Reasoning Mind. We began by asking a natural question: What exactly *is* the ideal implementation? Since that time, we have had the opportunity to observe thousands of them, visiting classrooms, speaking with teachers and students, and scrutinizing student learning data. Like the famous unhappy families of *Anna Karenina*'s opening line, unsuccessful implementations all fail in different ways. Nevertheless, there emerges a broad pattern. To achieve any measure of success, the classroom teacher must be a competent *facilitator*. And to rise above the merely adequate, and into the reaches of the unreservedly successful, the teacher must furthermore be a strong and active *instructor* (Figure 5.1).

What is entailed in these two roles? *Facilitation* is the administration of the classroom. As a facilitator, the teacher ensures that students are on task, are motivated, follow the instructions given in the online lessons, and have access to all of the materials they need—computers, notebooks, pencils, and so on. In a well-facilitated classroom, the students always know what is expected of them and adhere to it. No time is lost in finding seats and logging into the online program. No students sit idly for minutes, waiting for permission to use the bathroom. No frantic searches for erasers take place.

Facilitation	**Instruction**
• Keep students on task • Ensure students are motivated • Ensure students are following instructions given in online lessons • Provide access to all essential materials (e.g., computers, notebooks, pencils) • Establish efficient classroom procedures (e.g., for logging on, going to the bathroom, obtaining supplies) • Encourage productive peer interactions	• Ask students to *explain* mathematical ideas and problem solutions • Correct students' mathematical writing and speech • Coach students on mathematical problem solving

FIGURE 5.1. The activities of teachers in blended learning classrooms can be classified as either *facilitation* or *instruction*. Both are necessary for excellent teaching.

No students furtively while away the hours playing online pinball on a concealed browser tab. No calculators are used when they are not allowed. Besides these negative virtues, the classroom exhibits the positive virtues of good facilitation. Students are encouraged to try their best. They are recognized for their achievements. The mood is pleasant and upbeat. Students' questions are promptly answered. Students receive appropriate help from their peers. In other words, the classroom possesses what the corporate world calls "operating efficiency": the wheels turn as they should, with nothing neglected and nothing wasted.

But facilitation by itself is not enough for excellence. If a teacher is a brilliant facilitator but does nothing else, students only receive instruction from the online lessons. From the perspective of instruction modeling, this means that they only receive those experiences that can be automated; the entire host of instructional experiences that are essentially human is missing. But are these activities really all that important? If they are dispensable, our task is much simplified: just interview an expert mathematics teacher, build a computer program that mimics her teaching, sit the children down at computers, and train classroom teachers exclusively in facilitation.

So what are those elements of education that cannot be automated? Some of them fall into the facilitation category. Take for example the bond between the teacher and student: a good teacher relates to students, serving as a role model, mentor, and friend. The teacher provides encouragement, inspiration, and—when needed—a friendly kick in the pants. The teacher creates an atmosphere of excitement and wonder, which students come to associate with the subject. Ask anybody about their favorite subject from school, and invariably you will find a teacher whose force of personality made it so.

But beyond these facilitative tasks, there are also *instructional* tasks that cannot be automated. Take, for example, the matter of genuine understanding discussed in chapter 2. Students must constantly be asked to *explain* mathematics: to comment on their solutions to problems, to describe the reasons for their approaches, to discuss mathematical concepts. A good teacher has students do all of these things, coaching them to improve their reasoning and understanding. A second example is the matter of mathematical rigor. A good teacher corrects students' mathematical writing and speech, teaching them to express their thoughts clearly and precisely. And a third example is mathematical problem-solving. A good teacher coaches students through the messy process of tackling a new problem, guiding them as they learn to select, try, and discard different approaches until they find the key that unlocks the problem.

These activities are not peripheral to good mathematics instruction. As chapter 2 makes clear, they are what distinguishes outstanding from mediocre mathematics education. Today, they do not happen enough in American mathematics classrooms—partly because teachers do not have the time, and partly because they do not have the training—and no approach to improving mathematics education can be fully successful without remedying this problem. For this reason, instruction modeling is inherently *blended*: it must combine computer and face-to-face elements, carefully considering the role of the classroom teacher. Only in this way can instruction modeling succeed in its aim of bringing a complete world-class mathematics education to students.

This raises a natural question: if the human component is still indispensable, is there any point in going to all the trouble of instruction modeling in the first place? There are two main reasons why the answer is *yes*. First, there remain many important instructional activities that can be successfully automated: whole-class sessions (such as lectures), homework assignment and grading, independent problem-solving, skills practice, and much more. Automating these activities ensures that they are delivered with consistently high quality: students anywhere will have the same educational experiences during these activities that they would if they were in one of the world's best mathematics classrooms. Students are thereby guaranteed good content coverage, strong lesson plans, clear explanations, and well-chosen problems. And second, automating as much as possible frees classroom teachers to focus on the essentially human tasks. Now, teachers can have time to work with individual students or small groups, asking students to explain their reasoning, critiquing students' mathematical writing, and coaching students on problem-solving.

So it is not surprising that when a teacher is successful in *facilitation,* this makes it possible for her to attend to *instruction*. In an ideal Foundations implementation, the teacher has set up efficient classroom procedures and successfully promulgated them to students. The students know what to do without constant oversight from the teacher, allowing the teacher to spend her time on other things.

While good facilitation makes it possible for teachers to engage in instruction, it does not guarantee it. One observes a hierarchy of Foundations implementations (Figure 5.2). The worst implementations suffer from severe facilitation problems, with the teacher either disengaged or frantically attempting to control the classroom—leaving little time for anything else.

A step above are implementations with successful facilitation, but with limited teacher engagement in instruction. In these cases, teachers either do not realize the importance of instructional engagement, or lack the mathematical background to be comfortable with it.

Another step above, one finds classrooms where the teacher is instructionally engaged the majority of the time, but focuses exclusively on remediation: identifying and correcting basic gaps in students' knowledge. For example, after using online data to find that a student is struggling with a certain topic, the teacher can call him to her desk and tutor him individually. Foundations allows the teacher to pull up the specific problems the student has answered incorrectly,

FIGURE 5.2. The hierarchy of Foundations implementations.

so that the two can go over and correct the errors. The program offers complete, step-by-step solutions for every problem; after reviewing the solution and explaining the student's mistake, the teacher can generate a similar problem and ask the student to solve it himself. In some implementations, teachers spend all of their time on this sort of remediation. These implementations can often be very successful, but nevertheless fail to give students the essentially human instructional experiences which go beyond remediation.

Finally, at the peak of the hierarchy, one finds the rare, ideal implementations. Here, the teacher not only works on individual students' gaps, but also coaches students on solving advanced problems, frequently asks students to explain their reasoning in speech and in writing, and moderates meaningful mathematical conversations in small groups. Done properly, such implementations can provide the full richness of world-class mathematics instruction.

Resolving the Crisis of Implementation

The implementations of Foundations in the 2005–2006 school year had generally occupied the bottom two levels of this hierarchy in Figure 5.2. Our challenge—and it was an existential one—was to find a way to move implementations up. Moving teachers off of the bottom level was the most straightforward goal: since teachers already appreciate the importance of a well-run classroom, it suffices to give them the knowledge and skills needed to successfully facilitate a Reasoning Mind session. Moving beyond the second level was more difficult. Teachers first need to be persuaded of the value of their instructional engagement, and then given the knowledge of mathematics and the Reasoning Mind curriculum necessary to be comfortable and successful in that role. Even remedial tutoring requires good familiarity with the content: the teacher must not only be able to solve the problems himself, but also to find the root causes of student errors and offer clear explanations to correct students' misunderstandings. Moving to the fourth level is harder still. For example, moderating an open-ended discussion requires the mathematical knowledge to respond appropriately to a limitless range of student ideas, questions, and misconceptions. Even persuading some educators of the value of this sort of instruction is difficult, since one of the root causes of the problems of American mathematics education is a lack of appreciation for higher-order mathematical skills, such as mathematical writing, genuine understanding, and advanced problem-solving.

Our first step in addressing the challenge was to offer serious professional development for teachers. This began with a one-week, full-time course teachers would complete the summer before starting (called the Qualification Course), and a series of workshops they would do throughout the year (called the Certification Course). The Qualification Course focused primarily on facilitation, but also served to introduce teachers to the precepts of the Reasoning Mind curriculum, laying the groundwork for increased instructional engagement later. The Certification Course included equal coverage of facilitation and instruction. To make the importance of the training program clear to teachers, there were exams to ensure that all participants had mastered the material. Chief among them was the Qualification Exam, which teachers had to complete in order to pass the Qualification Course and be admitted to teaching with the program in the following year.

As we learned soon from the teachers, this approach was very unusual. Most programs were offered by for-profit "vendors," who naturally sought the path of least resistance. Extensive preparatory courses for teachers were uncommon, and mandatory exams more uncommon still. Many teachers had not taken an exam of any kind for many years, despite the frequency with which they inflicted them on their students. Despite all this, teachers responded well to the coursework. The trainings received high marks on anonymous surveys, and many teachers even reported liking the exam, as it demonstrated our commitment.

One of my favorite stories illustrating the new approach took place during one of the earliest Qualification Courses, in the summer of 2007. The course was taught by Steven Gaudino, the chief architect of Reasoning Mind's implementation department. Midway through the course, the teachers grew anxious about the upcoming exam. One of them raised his hand and asked, "What happens if we don't pass?" Steven had always believed in complete and unmitigated honesty. As a mathematics major, he also valued precision. He paused for half a minute to think over the question. Then he replied, pleasantly and matter-of-factly: "Well, there are three options. First, you could repeat the course next week and try to pass the exam then. Second, your school could send another teacher to pass the course. Or third, we don't work with your school." Taken aback, the teacher asked why the rules were so strict. Steven answered that they really were not so strict. We wanted to make sure teachers were equipped to use the program, and the exam merely checked this. If a teacher did not pass, then it was in no one's interest to proceed—not the school's, and certainly not the students'. This was hard to dispute.

Besides professional development, a team of *implementation coordinators* was put in place. Each teacher is assigned an implementation coordinator, who lives and works in the community she serves and is available to answer the teacher's questions by phone or e-mail throughout the year. The implementation coordinator visits the teacher several times a year, observing the class to give guidance on program use. They serve several purposes: to coach and advise teachers on the best ways to implement the program, to gather data on implementation quality across all of the schools and teachers using the program, and to persuade teachers of the value of following implementation recommendations (such as meeting the minimum time online, adhering to the recommended sequence of topics, teaching note-taking skills, being instructionally engaged, and so on).

One drawback of the implementation coordinator model of support is that it is expensive: adding up the time it takes to travel between schools, prepare for teacher meetings, and respond to teacher questions, it turns out that one implementation coordinator can only support several dozen teachers. A second drawback is that there is an inherent barrier when a representative of one organization (Reasoning Mind) gives coaching and suggestions to a representative of another (the school). A model in which school districts provided their own support to teachers would avoid both of these problems. However, such an arrangement at first proved elusive, since school district administrators (such as assistant principals, mathematics directors, and others) have many responsibilities, and merely adding this one did not allow them to devote the time and attention needed for success. Fortunately, as school districts expanded their implementations, it eventually became economically feasible for them to dedicate full-time (or close to it) administrators to overseeing the district's Reasoning Mind implementations. The first districts to do this successfully were Abilene Independent School District, a public school district in north Texas, and IDEA Public Schools, a network of high-performing public charter schools in several cities in the southern United States. Both of these organizations were able to establish and perfect the approach, leading to more economical and stronger implementations than was previously possible. Their work has become a model that other Reasoning Mind partner districts now work to emulate.

In the course of working to improve the quality of Reasoning Mind implementations, it has become clear that giving teachers the skills to be successful facilitators is an easier challenge than preparing them to succeed as instructors. Today, it is rare for implementations to languish on the bottom level of Figure 5.2. However, it is also relatively rare to find implementations at the very top one. In retrospect, this is not surprising: especially at the elementary school level, many teachers are uncomfortable with the subject. Only a few know it well enough to teach students more advanced material, such as how to solve multistep problems. We have found teachers to be aware of this gap in their knowledge and eager to remedy it, but the amount of content to be learned is very large and teachers have almost no spare time. The most favorable circumstances are to be found when teachers use Reasoning Mind as a core curriculum and are not responsible for teaching any other subjects, since then they can devote considerable time to professional development related to the program. For example, such teachers can complete "curriculum study," which means going through all of the online instructional materials as students. Of course, much still remains to be done to give teachers the best support in increasing their mathematical knowledge; we will return to this question in chapter 9, which discusses future directions.

Overall, the various improvements to implementation—professional development, implementation coordinators, and support models operated by districts themselves—have been successful in substantially improving the quality of implementations of the program. Evaluation studies again began to show positive results, as will be described in Chapter 6 (in the section on the program's effects).

The overarching lesson is that implementation matters as much as anything else. If badly implemented, a good program is indistinguishable from a bad one. This is true for any blended learning, but it is doubly true for an instruction modeling program. Why? Because the premise of instruction modeling is *instructional reform*: it is not merely about technologizing existing educational practices, but rather about changing the practices themselves. This applies as much to the *facilitation* as to the *instruction* provided by teachers. On the instruction side, teachers must learn to focus on teaching students to explain their reasoning, write clearly, express their mathematical ideas rigorously, and other such aspects of excellent mathematics instruction that cannot be provided by any machine. But the facilitation side, too, requires a change in mindset: teachers must emphasize the value of note-taking, prohibit the use of calculators when they are unnecessary, encourage students to work on more challenging problems, and communicate the importance of genuine understanding as students review the online materials. Instituting these practices requires close collaboration with teachers. To succeed, a program first must persuade teachers of the value of these reforms, and second must give them the knowledge and skills to enact them. In the human business of education there are no shortcuts.

While teacher training is the greatest practical challenge in instruction modeling, it is also one of its greatest strengths. A teacher in an instruction modeling classroom has access to all of the same online materials as students, giving him an excellent opportunity to learn the new curriculum. Moreover, the teacher comes into natural contact with the materials daily as he goes about his work. For example, a student might have a question, prompting the two to review an online lesson together; another student might struggle with a problem, pushing the teacher to understand the problem's solution so that it can be explained to the student; a third student might use the online curriculum's mathematical terminology during a small-group lesson, helping the teacher to acquire the new terminology himself. In other words, instruction modeling inherently *immerses* teachers in the tradition being modeled. This by itself may not be enough to give teachers all of the necessary knowledge and skills, but it is a great help.

In fact, some school systems have sought out instruction modeling expressly for its value in teacher development. One example is Great Hearts Academies, an Arizona-based network of several dozen public charter schools. Great Hearts uses a classical curriculum, complete with Latin for all students and classical Greek in the higher grades. Having toured their schools, I can confirm firsthand that they offer the sort of education one finds at the best private schools, but for free and with open admission to all who apply—an egalitarianism the ancients would surely have frowned on, but we moderns rightly appreciate. Great Hearts had chosen solid curricula for all of their subjects, including mathematics—they used a good series of foreign textbooks, adapted for use in the United States. However, they realized that their teachers were not mathematically prepared to deliver this curriculum, as they were mostly humanities majors with much intelligence and dedication but little training in mathematics. This led Great Hearts to implement Reasoning Mind.

I find this example particularly telling, since Great Hearts chose an instruction modeling program for its impact on teachers, not just for the underlying curriculum—and certainly not for the technology. In fact, the technology was at first difficult for them to accept, as they had an appropriate skepticism of educational technology, which so often offers ridiculous panaceas. Their approach stands in stark contrast to what usually happens: as Stanford historian Larry Cuban describes in *Teachers and Machines*, his brilliant history of twentieth-century education technologies, the main force leading schools to adopt technologies has been social pressure to be modern—regardless of the true instructional merit of the technologies themselves. There are right reasons and wrong reasons to use educational technologies, and the right ones are the curriculum and the impact on teachers' knowledge and practices. Instruction modeling is no exception: to succeed, it must attend not only to the computer-delivered lessons but also to the mathematical preparation of the classroom teacher.

Learning from Implementation: The Unexpected Lessons of Edifice

Foundations had taught us the importance of implementation. One might think that after many years of experience implementing this program, we would have learned everything there is to know. However, "implementation" is not a simple, monolithic notion: the ways in which students and teachers use different blended learning programs vary dramatically, and therefore the challenges of implementation vary as well.

A case study for this is the Edifice program. This program—including a complete core sixth grade curriculum—was finished in the summer of 2013 and piloted in three schools. It has been in continuous use since then, with a core seventh grade curriculum added in 2015 and a plan in place to eventually add an eighth grade, making it a complete middle school program. Because the program was such a dramatic departure from Reasoning Mind's previous work, it is in retrospect no surprise that much was learned when the program entered classrooms.

One major difference between Edifice and Foundations was in the structure of the lessons. In Foundations, a "lesson" (technically, they are called "objectives") could last anywhere from thirty minutes to three hours; a student can complete a lesson over several days, each day continuing from the point where he last stopped. This approach is a failure of modeling, since in their regular work, the expert Russian mathematics teachers plan each lesson to be a self-contained story. The teachers describe each lesson almost as if it were a play, with each portion leading into the next to ultimately form a unified, inseparable whole. To improve the fidelity of modeling in Edifice, we decided to follow this practice: a course would consist of one hundred twenty lessons (or thereabouts), with each meant to be completed in one sitting. If a student failed to complete a lesson, she would either need to start it over the next day, or (if she had gotten far enough) would be bumped to the next one.

This change, initially made in the name of more accurate modeling, had some interesting unintended consequences. Because Foundations was entirely self-paced, students were at a wide range of points in the course at any given time. This meant that some invariably did not complete the entire course. Furthermore, the teacher could not plan whole-group offline sessions that were well coordinated with the online materials, since only a fraction of the students would be on the same topic at the same time.

By contrast, Edifice's requirement that each lesson be completed in one sitting forced the class to move through the course roughly at the same pace. Besides giving teachers the comfort of knowing that all students would—one way or another—complete the course, this gave teachers the ability to tie their whole-group instruction into the online lessons. For example, some Edifice teachers would begin each day with a ten-minute preview of the day's material, or else conclude the day with a group discussion of what was learned. This also pushed teachers to learn the material better themselves, since now the content of every lesson was immediately relevant to their plan for the day.

But there was a problem with requiring students to complete one lesson every day: because students still chose their pace of progress *within* the lesson, some students would end up taking far longer than average, thus having to either repeat lessons or skip important material as they were pushed on to keep up with the class. We studied the data and found one cause that accounted for almost the whole problem: when students were asked a question they could not answer, they would sometimes take far longer than was constructive. For example, one question asked students to divide two numbers with remainder; some students who had forgotten how to do this would stay on the screen for ten or fifteen minutes, scribbling unproductively in their notebooks, before finally giving up. It is one thing to take time to think about a challenging problem, but it is another thing altogether to spend ten minutes working on something fruitlessly when the barrier is not a lack of thinking, but rather a lack of some specific piece of knowledge. In such cases, it is better for students to quickly admit that they do not know, so that the teacher can provide an explanation. In a live class, this is exactly what happens: if the teacher calls on a student, she will not stop the class for ten minutes while the student thinks. After several seconds of silence, she will call on someone else or provide the answer herself.

To mimic this behavior in Edifice, we added timers to every question. If a student tarried longer than the question merited, the tutor character would gently remind him to wrap up his work; if he did not answer in a further ten seconds, the question would be marked as incorrect and the lesson would move on. While the timers were set quite liberally, they nevertheless served to focus the students, who knew they had to work efficiently on every problem presented in the lesson.

Between the timers and the requirement that all students work on the same lesson each day, Edifice is clearly less self-paced than Foundations. Does this mean it is worse? After all, "self-paced" is a favorite buzzword among the promoters of blended learning. Fortunately, this is precisely the sort of question that can be—and, in fact, has been—tested in the laboratory. The research case for self-pacing is that students fare better when they have the necessary time to absorb each piece of information before having to move on; otherwise, they end up trying to process the previous and next idea at the same time, overloading their working memory. This is the substance of the cognitive theory of multimedia learning (CTML)'s *segmenting principle*,[1] which states that material should be given in small bits, giving students a chance to understand each one before moving on (see chapter 4 for more detail on the CTML). Understanding now the logic behind the segmenting principle, we see that Edifice does not violate it: students *do* have the opportunity to process material at their own pace, except in cases when they are stuck on a question and are almost certainly no longer being productive. Of course, there is also the separate question of which approach students *like* better, and the implementations of Edifice have shown that the timers, while tolerable, are not a hit with students. But we do not always enjoy what is best for us.

Their popularity aside, the timers—in combination with headphones (which were mandatory in Edifice, unlike Foundations, because of the sound)—had the unintended effect of engrossing students in their online lessons. The atmosphere in an Edifice classroom was strikingly different from a Foundations classroom. Foundations classrooms already exhibited unusually high time on task, but in Edifice it was nearly absolute. There was no small talk between students. The facial expressions of students were all the same: the blank stare of concentration. The teacher would not make intermittent comments, since the students could not hear them. Walking through an Edifice classroom was a strange experience: the complete silence of thirty students working intensely at their desks.

It was interesting to see the vastly different reactions of visitors to Edifice classrooms. Those who were experienced in education—such as teachers or principals—would see a group of focused students, hard at work on the task of learning mathematics. Meanwhile, members of the laity would see an educational dystopia, a cold and impersonal factory of knowledge, with thirty students staring in silence at their screens. I myself side with teachers in this question. We are fed images of the perfect classroom: an active, bustling room; educational brochures are decorated with pictures of laughing students, sitting in small groups or working on "hands-on" projects. But learning does not happen through hustle or through bustle—it takes place in the mind. It requires focus and freedom from distraction. Of course, conversation plays an essential role: students must ask questions, discuss what they are learning, and explain their reasoning. But this is only needed for some portion of the time: the majority should be spent in silent work, whether listening, reading, or independently solving mathematical problems.[2] The ultimate determinant of learning is not the apparent liveliness of the classroom but the substance of the educational experiences of the students. Teachers, unlike the uninitiated, understand this. I am reminded of a time when Tom Torkelson, founder and chief executive of IDEA Public Schools, visited an Edifice classroom. Instead of walking around the room looking out of place—as most visitors do—Tom found a student using earbuds and asked her to split them. Then Tom sat for fifteen minutes, listening in as the student went through the lesson so that he could follow and understand her experiences. An uninformed observer might have thought they were listening to a new Justin Bieber album together.

The engrossment of students had a surprising effect on classroom teachers in Edifice. In Foundations, teachers could spend their time primarily on facilitation—in what Figure 5.2 called Level 2 implementations—without feeling idle. They could walk around the room, answer student questions, and occasionally offer a hint when a student appeared to be struggling. While this felt productive to teachers, it was in fact wasteful: the teacher's time would have been far better spent working with individual students or small groups, remedying gaps in prior knowledge or asking students to explain their solutions to problems. In Edifice, the picture was completely different. With students wearing headphones and on timers, there was no room for incidental hints from the teacher, nor were students tempted to delay their work by raising their hands to ask unessential questions. For the teacher, walking around the room was now an obviously useless activity. This pushed teachers to spend their time primarily on instruction, pulling individual students off their computers and working with them in separate sessions. It also pushed teachers to spend part of each day on whole-group instruction tied into the online lessons. Thus, yet another unintended consequence of the timers and headphones emerged: pushing Level 2 implementations to rise to Level 3.

Why Comprehensive Programs Spread Slowly

Edifice also serves as a case study for another general rule: because comprehensive programs* are harder to implement, they spread more slowly. We had already seen this with Foundations: the fifth grade, which was a core curriculum (i.e., intended to fully replace ordinary instruction and containing over a hundred hours of online lessons), expanded to more users substantially slower than the second, third, and fourth grades, which were ancillary (i.e., intended to be used for several dozen hours over the course of the year, alongside the regular curriculum used by a school). Because Edifice was exclusively a core curriculum, it was in the same boat as the Foundations fifth grade.

The difficulty with establishing core curriculum implementations was not that individual teachers, principals, or district administrators did not like the idea of blended learning as a core. Rather, it was the consequence of a variety of logistical and bureaucratic barriers: a school needed more computers to use the program as a core; more internal approvals were needed, including from various departments in the school district central office; there was a fear that any change to the core curriculum (especially one as radical as moving to a digital

* Instruction modeling programs, which aim to replicate some system of instruction, naturally tend to be comprehensive.

program) would jeopardize test scores; more teacher professional development was needed, adding further scheduling challenges.

Interestingly, once a school district *did* choose to use the program as a core curriculum—whether with Foundations or Edifice—the average quality of implementation was higher than with ancillary programs, since teachers and administrators assigned the program a higher priority and were able to dedicate more time and attention toward its successful use. However, this did not solve the problem of getting them there in the first place. With Edifice, the challenge was even greater than with the fifth grade core program in Foundations, since Edifice requires each child to have headphones, has higher bandwidth needs (due to the greater amount of sound), and requires enough time to be scheduled for each session to ensure completion of a whole lesson. These aspects of the program were instructional strengths, but also made it less flexible.

The challenge is not unique to Reasoning Mind. In 2016, researchers at SRI International analyzed usage information from several dozen education technology programs at the college level; they found that the more comprehensive a program was, the fewer users it had.[3] This is disheartening, since a program cannot be effective if it is not used for many hours (even a hundred hours is a small length of time where instruction is concerned) and offered along with extensive training and implementation support for teachers. In other words: to be effective, a program must be comprehensive, but getting any comprehensive program to be implemented is extremely difficult.

At this point, a natural question arises. If comprehensive programs are better, would not the market push schools to adopt them in spite of the effort they require? Would not the invisible hand of Adam Smith, gentle but inexorable, guide us toward mathematical excellence? The answer, unfortunately, is *no*. This is not because the market is absent or fails: traditional public schools *do* face market pressures, as they increasingly compete against public charter schools for students and teachers. They also face political pressures, as parents advocate with school board members for policies they believe will benefit their children. However, these market and political pressures do not directly reward instructional quality. This is because unlike corporate profit, instructional quality does not come attached to a number. The proxy that governments and parents alike have chosen is *test scores*. But, as we saw in the first chapter, test scores can only measure instructional quality when they are used in the context of carefully designed experiments or statistical analyses, and even then contribute only a part of the evidence needed to spot good instruction. Over a short period of time (such as one year), an excellent curriculum will *not* make a large enough impact on test scores to be visible to the naked eye. What *will* make such a difference, however, is the *alignment* of the curriculum to the test; this is the *nefarious alignment effect* discussed in the first chapter. Thus, schools direct their energies toward ensuring that their curricula are excellently aligned to the tests: exactly what the market and political pressures demand, but largely irrelevant to instructional quality. When teachers and principals select a program that is instructionally excellent but difficult to implement, they do so not because of market pressures, but in spite of them.

This last point cannot be stressed enough. Because so many members of the general public—and even many educators—erroneously think test scores are a far more accurate measure of learning than they are, schools are forced to direct their energies toward bureaucratic compliance over genuine education. Simply doing away with tests would not be a solution, since the absence of false knowledge does not by itself confer the correct variety. The only remedy, ironically, is education: not of children, but of us adults. If we care about the quality of our schools, then it is our civic duty to learn how to tell apart good from bad instruction, with all of the complexities that it entails.

Lighter Fare: Smarter Solving, STAAR Readiness, and Blueprint

Comprehensive programs are the most effective, but also the most difficult to implement. For Reasoning Mind, this created a problem: having only comprehensive programs to offer to school districts, any partnership instantly required a high degree of commitment and trust. Furthermore, expanding to more classrooms *within* a school district was always a slow and difficult process. A nonprofit organization like Reasoning Mind does not have pressure to produce profit, but must nevertheless remain solvent. Rapid growth is essential, both because revenue from tuition is needed to cover costs and because philanthropic donors expect it as a precondition for continued funding. And of course, growth is important in itself, since the motivation for instruction modeling is that it can bring excellent instruction to *many* students rather than just to a lucky few.

Several years ago, a philanthropic foundation paid for a team of management consultants to advise Reasoning Mind on growth. They concluded that to grow at the desired pace, the organization needed to allow the fifth-grade program—intended for use as a core curriculum—to be used as an ancillary (i.e., noncore) curriculum, for the implementation support to be scaled back, and for topics to be assignable in any order (whether mathematically logical or not). It would amount to gutting the program, making it more appealing to schools at the expense of instructional quality. The consultants were completely right: the program as it stood was far too comprehensive—and consequently inflexible—to expand to more schools at the required rate.

It seemed to be a problem with no solution, on one hand the Scylla of stagnation, on the other hand the Charybdis of cynicism. Monetary bankruptcy or instructional bankruptcy. At the time, we chose to do nothing. To our great luck, a solution eventually fell in our laps. Schools had complained that Reasoning Mind's programs completely ignored the requirements of state high-stakes tests (such as the Texas STAAR test), neglecting some topics covered on these tests and presenting no problems with similar style and phrasing to those that appeared on the tests. To address this concern, we developed a collection of "STAAR Readiness" lessons, designed to brush up on topics covered on the Texas exam. Because they were not meant to be the principal vehicles of instruction, these lessons were brief and could be assigned in any order. Schools loved them. More surprisingly, teachers and administrators perceived them as separate from the rest of the program. This led to a natural idea: in cases when a school was not ready for the commitment entailed in a comprehensive implementation, why not offer the test readiness lessons on their own?

This idea became a strategy. Reasoning Mind would maintain a variety of different programs, some comprehensive and others far lighter. Partnerships with school districts would usually begin with the lighter programs, which would more obviously meet districts' immediate needs. Of course, these programs would not be instructionally transformative—i.e., they could never make up the difference between mediocre and excellent instruction—but would still be helpful, bringing some benefit to students. Meanwhile, as partner districts became more comfortable with Reasoning Mind and blended learning, they would eventually move to the more comprehensive programs, including core curriculum. So far, the strategy has been a success: many new districts have signed on to the lighter programs which would not have been ready yet for the comprehensive ones. Just as importantly, districts that started with the lighter programs have shown a willingness to move up to comprehensive programs in subsequent years. The lighter programs have become to the comprehensive ones what dating is to marriage: not a substitute, but rather a pathway.

Three of these lighter programs were eventually developed: STAAR Readiness, Smarter Solving, and Blueprint. STAAR Readiness grew directly out of the lessons we had already made, and was simply a program that offered these lessons on their own. Because the lessons were part of Foundations, the STAAR Readiness program uses the same interface and approach (Figure 5.3). The lessons include both explanations of concepts and practice problems.

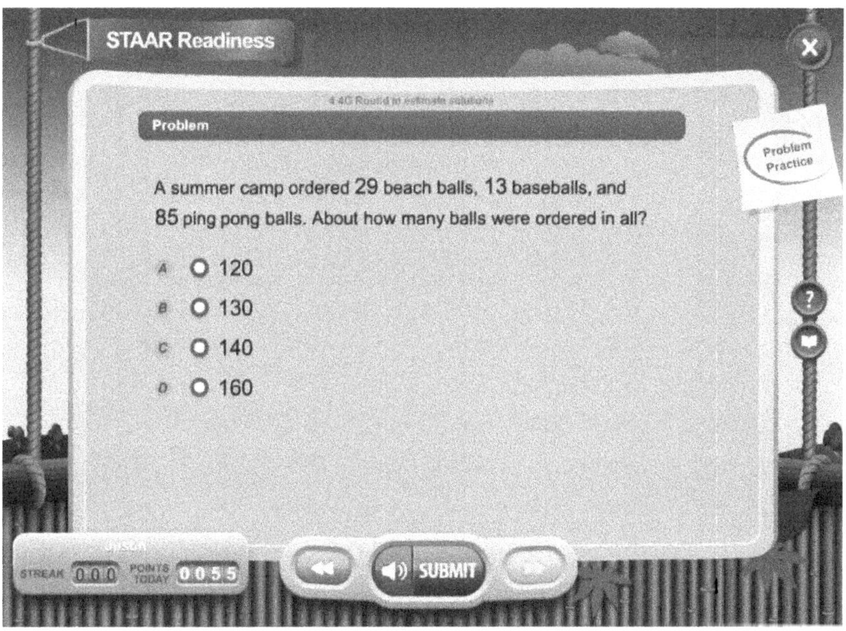

FIGURE 5.3. The STAAR Readiness program.

The Smarter Solving program is designed to develop advanced problem-solving skills. Schools are especially concerned about this in light of the more difficult problems offered on new high-stakes tests, such as the Smarter Balanced and the PARCC assessments. However, it is an important instructional aim in itself, since—as chapter 2 explained—genuine mathematical problem-solving is one of the central differences between outstanding and mediocre mathematics instruction. Using a similar interface to Foundations, the Smarter Solving program presents a wide variety of problems, each requiring students to use a different combination of strategies (Figure 5.4). Starting with simpler problems and gradually building up to more complex ones, the program teaches students to apply the underlying strategies in varied combinations and contexts. Each problem includes a detailed, step-by-step solution, so that students who struggle with a problem are shown how to approach it.

The decision to use an interface similar to Foundations—rather than Edifice—was deliberate in Smarter Solving. Since the program was based primarily around problems, rather than discussions or explanations, that interface was sufficient to provide many of

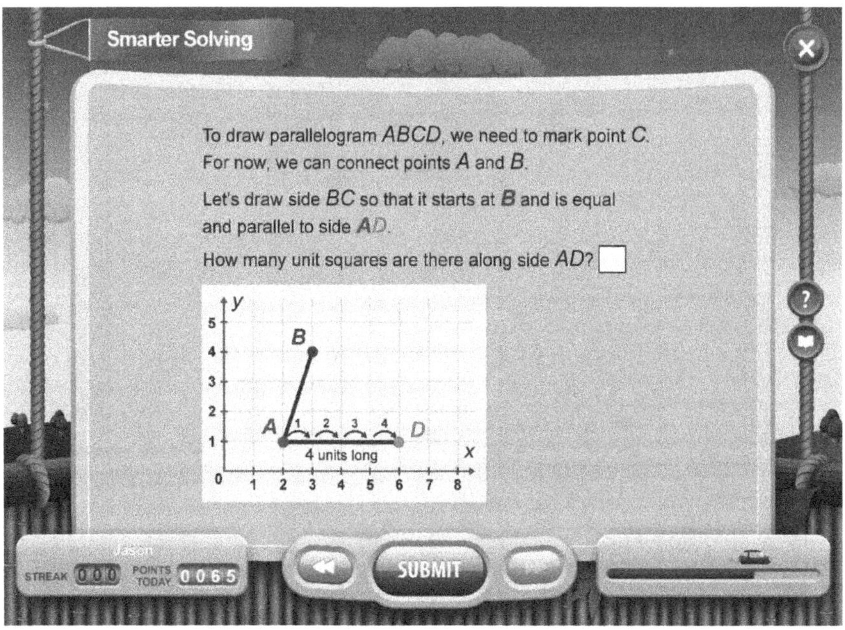

FIGURE 5.4. The Smarter Solving program.

the necessary experiences to students. Just as importantly, creating materials for Foundations is substantially less laborious than for Edifice. And finally, lifting the requirement of headphones made a program more flexible and easier for schools to adopt, which was the primary motivation behind the lighter programs in the first place. This illustrates some of the trade-offs in selecting an interface. A more robust learner interface is not always the best choice: in some situations, a simpler interface may be sufficiently instructionally effective with fewer barriers to implementation.

To what extent do STAAR Readiness and Smarter Solving rely on instruction modeling? Beyond a doubt, less heavily than Foundations or Edifice: the expert Russian mathematics teachers do not have direct prior experience preparing students for the STAAR exam, or developing problem-solving skills in isolation from the rest of the mathematics course. However, they nevertheless have experience preparing students for *other* exams, and also experience developing problem-solving skills as part of their regular work. The materials they prepared for these tasks were clearly superior to what our team could have done otherwise, with the difference especially obvious in the case of problem-solving skills. Thus, we chose to use the same process as in Foundations, with the expert teachers writing the initial versions of the materials and then the members of the curriculum development team editing them to produce a final version. After all, instruction modeling is not a *process*, but rather an *approach*. It can be applied in many different ways and to varying extents. The only invariant is the attempt to reproduce the experiences of students in the classrooms of exceptional teachers, thereby bringing the benefit of these teachers' knowledge to more students.

The third of the lighter programs, Blueprint (Figure 5.5), is intended for students in prekindergarten, kindergarten, and first grade. These grades are unique. First, children of this age do not require a continuous hour of mathematics instruction daily: two daily lessons of twenty minutes each should suffice, and even the most ardent lover of mathematics would agree that much more would constitute an overdose. And second, children at this age are not subject to high-stakes testing, which usually begins only from the third grade. (Apparently, the higher powers have judged second graders too fragile to endure testing, but found third graders to be just right.) Thus, schools are substantially less anxious about changing

FIGURE 5.5. The Blueprint program.

their instructional practices at these grades, and are better able to make decisions purely on the basis of instructional quality. These two circumstances allow a program to be simultaneously *light* and *comprehensive*, something impossible in higher grades. Blueprint, the brainchild of my mother, Julia Khachatryan, takes advantage of this special situation.

The program consists of one hundred lessons per year, each lasting twenty minutes. The content of the lessons is obtained from the expert Russian mathematics teachers. To save development time, this is done in a manner between that of Edifice and Foundations: the teachers write lessons as sequences of assignments and activities, neither referring to the computer nor explicitly documenting their offline practices; then, a knowledge engineer determines how to model these activities through online lessons. Meanwhile, the interface of Blueprint is more like Edifice. While it is not a simulation of a group lesson, it mirrors Edifice in closely following the recommendations of the cognitive theory of multimedia learning, including separation of voice and text, informal tone taken by the narrator, and gesturing to draw the students' attention to graphical elements when they are mentioned by voice. To be suitable for very small children, the program embeds the instruction in a make-believe world, where the student learns alongside human and animal characters.

The three lighter programs—STAAR Readiness, Smarter Solving, and Blueprint—illustrate the value of using varied interfaces and taking varied approaches to instruction modeling. There is not a simple, linear path from worse interfaces to better ones, or from less faithful to more faithful modeling. Some interfaces are certainly more cognitively robust, and some approaches to modeling are surely more meticulous, but these improvements could carry greater cost, harder implementation, or both. Trade-offs are inevitable.

Chapter 6

The Case for Instruction Modeling

THE PRECEDING THREE CHAPTERS PRESENTED THE CORE ideas of instruction modeling: what it means, how it relates to learning science, and what implications it has for implementation. Now, it is time to reflect. Instruction modeling is obviously a laborious process. Why is it worth the effort?

The Three Dimensions of Quality in Blended Learning: Content, Interface, Implementation

Before we can discuss the value of instruction modeling, we should answer a broader question: what qualities does a blended learning program need to possess in order to be effective? Then, if we find that instruction modeling confers one or more of these qualities, this will be an argument in its favor. If, however, we find that instruction modeling is irrelevant to these qualities, then we can conclude that it is not worth the effort.

I have found that the most useful way to think about any blended learning program is to break it down into three distinct dimensions: the *content*, the *interface*, and the *implementation*

Instruction Modeling. George A. Khachatryan, Oxford University Press (2020)

DOI: 10.1093/oso/9780190910709.001.0001

(Figure 6.1). We have already encountered each of these three notions in the preceding chapters, but this is nevertheless a good opportunity to review their meaning.

The *content* is the substance of the instruction. It consists of all of the information that is communicated to students and all of the tasks they are expected to perform. The content includes the problems that are assigned, the explanations that are given, the examples that are offered, and so on. It is independent of the manner in which these things are delivered; for example, no matter whether a problem is given on a worksheet, on a computer screen, or verbally by the teacher, the *content* remains the same: it is the problem itself.

To further illustrate what makes up the content, consider the example of *work at the board*, the activity in which the teacher calls on a student to present a solution to the class, writing the steps on the board and solving the problem through a collaborative conversation with the teacher. This activity offers very helpful *content* to the students at their desks: they see an example of how to reason through a problem solution. An online lesson that included all of this dialogue would preserve this content. Meanwhile, an online lesson that merely presented the problem with a brief finished solution would not. Thus, the *content* entails *all* of the information that is present in a lesson, not just the headlines.

Meanwhile, the *interface* is the particular way in which the content is delivered. The word "interface" is popularly used in reference to the user interfaces of software programs; here, however, I mean it in a broader sense to encompass *all* of the interactions students have, whether online or in person. It certainly includes the user interface

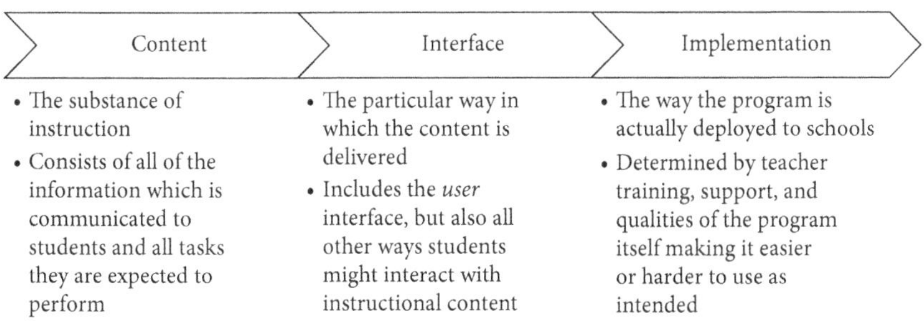

FIGURE 6.1. The three dimensions of effective blended learning.

of any online lessons students might take, but it also includes all of the *other* ways students might interact with instructional content. For example, an explanation of a mathematical idea might be delivered through text on a screen, through an online video, through words in a printed book, or through speech by the teacher. Each of these is a different *interface* delivering the exact same content. As we saw in chapter 4, the interface plays an important role in determining the student experience. A badly chosen interface can distract students from the content. Or, it could overwhelm them (as in the case of screens full of text in online lessons), preventing students from actually attending to the content. And even among those interfaces that *do* lead students to attend to the content, some encourage active thought and understanding better than others. We saw an example of this in the *personalization principle* of the cognitive theory of multimedia learning, which posits that information is better absorbed when it is offered in a way that successfully emulates a real social interaction—in particular, is delivered by a speaker who is personified, friendly, and uses a real (rather than synthesized) voice.[1]

Finally, the *implementation* is the way in which the program is actually deployed to schools. It is determined by the training provided to teachers, the support they are given throughout the school year, and the qualities of the program itself that make it either easier or more difficult to use as intended. The implementation affects the form in which the intended content and interface actually reach students. For example, imagine a blended learning program which presupposes that while students work online, the teacher spends time one-on-one with those students who are struggling. The intended interface for these struggling students is a rich one indeed: they have the benefit of a true two-directional conversation with a knowledgeable adult! However, if an implementation fails to motivate teachers to work directly with students, then this activity will never take place. The wonderful interface *intended* by the program will remain an intent only. This is why Steven Gaudino, Reasoning Mind's vice president responsible for program delivery, likes to say that implementation is the craft of turning the imagined into the real.

These three dimensions fit neatly together, collectively offering an exhaustive description of any blended learning program. The content is the substance of the instruction. The interface is the way in which that substance is delivered. Together, they make up the *intended* program. The implementation determines whether the *reality* of the program resembles this intent.

For a blended learning program to be effective, it must be successful in *each* of the three dimensions. If the content is rotten, then a program cannot possibly be excellent, no matter how well this rotten content is delivered. If the content is excellent but the interface is badly chosen, then students will either ignore the content completely, or else process it superficially and without active thought; the content in this case *might as well* have been rotten, since no one will have noticed it anyway. Finally, if the implementation is weak, then the most beautifully designed program—with wondrous content and a magnificent interface—will be of no worth, since it will be used so little as to be irrelevant—or, at best, used much but so badly as to be unrecognizable.

These three dimensions in themselves are nothing surprising or groundbreaking. In fact, they are merely a *perspective*: one particular way to decompose any blended learning program into aspects that can then be studied separately. There is an infinite number of other ways to do this, and no one way is "correct." However, some ways are certainly more *illuminating* than others. I have found this particular one to be especially valuable, and believe that many who have been deeply involved in blended learning would agree.

How do most of today's blended learning programs fare on these three dimensions? We will work backward, starting with *implementation*. An excellent source for this is *Teachers and Machines: The Classroom Use of Technology Since 1920*,[2] a book by Stanford historian Larry Cuban that should be required reading for anyone involved in blended learning. Many assume that education technology is a recent invention. In fact, every generation of the last hundred years has had its own instructional technologies: instructional film in the 1920s, radio in the 1930s, television in the 1950s, and "computer-assisted instruction"—the worse-branded predecessor of blended learning—in the 1980s. Cuban chronicles these waves, and the similarities that emerge are remarkable. In each case, implementation was the programs' undoing. The programs were not designed with an understanding of teachers' real needs and the realities of school life. They were offered with insufficient training and support. And, ultimately, they were used sparingly and as diversions instead of as the wholesale reforms that the developers had envisioned. Cuban carries out a telling—though somewhat scatological—calculation: in each wave, students ended up spending less time with the technologies than they spent going to and from the bathroom.

Today's technologies follow in the tradition. Comprehensive teacher training is almost unheard of, as is implementation support. Usage of adopted programs is low. For example, a prominent blended learning program recently underwent a large research study to assess its effectiveness; the study, likely out of a desire to be agreeable, praised the program in the summary and buried the lede: over the course of a full year, participating students had spent fewer than twenty hours actually using the program. What can be expected from twenty hours of instruction in a discipline as large as mathematics?

Fortunately, there are glimmers of hope. Implementation is receiving increased attention. For example, Michael Horn and Heather Straker have made it the focus of *Blended*,[3] their bestselling book. Frequent speakers on blended learning, Horn and Straker have done much to make school districts and their broader communities aware of the challenge. But schools have yet to demand implementation support as a precondition to adopting a program, and they have yet to accept the need to fund such support in cases when it is offered. Program developers, taking their cue from schools, have largely not developed adequate implementation services. In other words, much work remains to be done, but more people are becoming aware of the problem—and this is a promising trend.

Turning now to *interface*, the situation here is somewhat better. Education psychologists like Richard Mayer have built a large body of scientific evidence to identify the principles of good design. Some developers of blended learning are well- aware of these principles, and apply them consistently in their work. An outstanding example is Kaplan, the test preparation company. Under the leadership of Kaplan's chief learning officer, Bror Saxberg, the company has promulgated a list of research-based practices for designing online lessons. All of the company's business units are urged to follow the practices, and aided in this effort by Saxberg's team of instructional designers.[4] However, not all publishers have followed Kaplan's lead. Academic conferences of educational psychologists are poorly attended by program developers, and psychologists often lament the lack of interest that publishers take in applying their work. As a consequence, there is great variance in the quality of program interfaces: some programs follow the principles of good interface design closely, but many others take no heed. Considering the interface in the broad sense—in particular, to include interactions with the classroom teacher—few programs have provisions for encouraging this. Until schools acquire a deep understanding of the relevant recommendations of learning science and begin to expect conformance from providers, the situation is unlikely to change much. But the more conscientious developers are already taking positive steps, regardless of the fact that this may not increase schools' demand for their programs.

Finally, we arrive at *content*. Here the situation is dire. I discussed in chapter 2 the low quality of American mathematics textbooks. Blended learning programs are similar. If anything, their content is worse, as the attention of developers is directed toward the capabilities of the technologies over the mathematical substance that underlies them. Of course, there are some exceptions. For example, in developing its *Cognitive Tutor* program, Carnegie Learning paid careful attention to the content, aiming to provide conceptual explanations of mathematical concepts, building on previous knowledge as students moved through the program. When this conflicted with the approach in the printed textbooks schools used, the developers went so far as to write and publish their own textbooks.[5] Unfortunately, this sort of attention to the content is uncommon.

A mathematician friend of mine now works for a company that analyzes large volumes of online student data to recommend lessons from a large collection. The idea is that anyone can develop their own lessons, and then the company's algorithms will identify the best. My friend confided in me that there was a problem: there were almost no lessons with good content to be found in the first place. Without good content, we can be as technological as we please; it will make no difference.

To summarize the state of things, today's blended learning programs suffer in implementation, but are beginning to take notice; vary widely in interface; and are near-universally rubbish in content. If I were to grade leniently, I might award a C for implementation and a B for interface. For content I would have no choice besides a failing grade.

Why Instruction Modeling Is Worth the Effort

What do we seek from blended learning? If we seek a toy, a pleasant diversion from the strain of actual learning, then we can comfortably disregard the three dimensions of quality discussed in the preceding section. As long as a program is colorful and entertaining, and carries the outward appearance of something that might plausibly confer knowledge, we would be pleased. Indeed, an excessive preoccupation with the quality of a program that is only used for a dozen hours over the course of a year would indicate a mind dangerously detached from reality.

However, many of the instructional designers who build blended learning programs, many of the teachers who use them, many of the academics who study them, and many of the enthusiastic bystanders who cheer them on hope for more. At the extremes one hears talk of a coming revolution in learning. More cautious proponents avoid the language of the guillotine but nevertheless speak of a substantial improvement in instruction over what was previously possible. Will any of this come to pass? Only if the quality of available programs increases along all three dimensions: content, interface, implementation. We cannot hope to achieve good instruction with bad programs.

What must be done to ensure that programs are strong in each dimension? Looking after the *interface* and the *implementation* is not easy, but is still relatively straightforward: to achieve a minimum acceptable level* of implementation, it is enough to create adequate training and support programs for teachers and then deploy them consistently; to design strong interfaces, it is sufficient to understand a few dozen ideas from learning science and apply them correctly. In other words, developers need to make a relatively small number of decisions well at the outset, and then consistently stick to these decisions from that point on.

Adequately addressing *content* is far more difficult. As we saw in chapter 2, instructional quality resides in the details. Good content must certainly follow a set of principles, but applying these principles well is a great challenge—and the problem must be solved anew

* Achieving truly outstanding implementation is much more difficult, since it requires providing all participating teachers with deep knowledge of the content and how to teach it.

for every minute of instruction. In other words, the creation of good content requires developers to make a massive number of decisions well. And these decisions are exceptionally difficult ones: after all, they are exactly the same ones that textbook authors and mathematics teachers must make in their work, and the paucity of countries with outstanding mathematics education proves beyond any reasonable doubt the difficulty of the task.

Good traditions of mathematics instruction are rare, and were created through generations of effort by remarkable mathematics teachers, mathematicians, and psychologists. A developer of blended learning programs who wishes for good content faces nothing less than the task of matching the quality of these traditions in every lesson, screen, problem, animation, and question. Developers who flippantly believe they can achieve this without a systematic, deeply considered plan are victims of hubris. Developing excellent blended learning content is the same fundamental challenge as delivering excellent mathematics teaching, and requires the same massive amount of rare, specialized knowledge. This—and nothing else—is the greatest hurdle for building effective blended learning programs.

This challenge is the primary motivation for instruction modeling. *The value of instruction modeling is that it provides a consistent, reliable way to obtain high-quality content.*

There are three special virtues of instruction modeling that make this possible. The first is that instruction modeling ensures that the developed content attends to *all* of the aspects of good education, not just the ones that are easily measured. In the first chapter, I drew a distinction between quantitative and humanistic ways of thinking about education, arguing that each offered something the other did not. Quantitative methods offer greater certainty of knowledge, but methods from the humanities can teach us about the less tangible—but still essential—qualities of good instruction. One form of humanistic inquiry is ethnography, and instruction modeling is an ethnographic approach: it does not happen in the laboratory, but through the careful study of existing traditions of mathematics education. And this is a great virtue, allowing instruction modeling to attend to the intangibles of good education. Whatever takes place in our chosen tradition to delight and inspire students, to impart to them ideas of mathematical beauty, or to mold their thinking—all of these things will be captured and replicated through the content of well-done instruction modeling.

This is especially important for blended learning. It is easy to get caught up in the apparent benefits of technology: we hear that it provides detailed data on students' performance, that it can adjust to each student's individual needs, that it can engage students with colorful graphics . . . but is this enough? Good instructional content goes far beyond these things, and we ignore this at our peril. Instruction modeling forces us to take a comprehensive approach. The instruction modeler asks not, "Does this have some features I like?" but rather, "Does this content include *all* of the things that make for excellent instruction?" In other words, instruction modeling ensures a holistic view and prevents us from unintentionally and unwittingly offering an impoverished education.

A second virtue of instruction modeling is that it is a fundamentally cautious approach. Many education technologies aim to design the best education possible, hoping that computers can offer better education than previous generations of students could ever experience. By contrast, the instruction modeler sets a modest goal: to take currently existing high-quality instruction, and to use computers to make its content available to many more students and teachers. The instruction modeler does not seek the new, but only the increased availability of the best of the old.

This may seem disappointingly pedestrian. But we know already that outstanding traditions of mathematics education are rare. And the ones that do exist were created over decades or even centuries, slowly refined and perfected by successive generations. Any attempt to start from scratch—no matter how thoughtful or well-informed—is most likely to end in disappointment, just as the vast majority of past attempts have ended. If our goal is to provide excellent instructional content, our best bet is to *begin* with content we already know to be excellent.

Finally, the third major virtue of instruction modeling is that it makes sure that the content of what is being developed is good not only in its broad outlines, but in all of the details that make up each lesson. The same thing that makes instruction modeling so laborious—namely, the need to model each lesson individually—ensures that each piece of the result is of good quality. And, as we saw earlier, this is the hardest thing to achieve in developing any educational program. Picking a good collection of guiding principles is one thing, but providing consistently excellent instruction is another thing altogether. Without a safeguard in place to ensure the consistent quality of every minute of instruction, the result is sure to fall short of excellence. Instruction modeling provides this safeguard.

So far, I have spoken only of the value of instruction modeling in producing good *content*. But what about the other two dimensions, *interface* and *implementation*? Instruction modeling forces developers to constantly compare their programs with the best of offline instruction, and this has a salutary effect on implementation. For example, in Reasoning Mind's case, this comparison helped to strengthen the argument for the importance of instructional involvement by classroom teachers: in classes where the teacher acted as a facilitator but not as a one-on-one or small-group instructor, it was clear that students were receiving a less rich educational experience than in the high-quality offline classes Reasoning Mind was seeking to emulate. This in turn pushed the organization to do more to encourage teachers to become more instructionally involved. However, all told, instruction modeling played only a supporting role; the main forces driving the development of Reasoning Mind's implementation practices were honest observation and a commitment to improvement. For this reason, when it comes to *implementation*, instruction modeling is helpful but not irreplaceable.

Of course, instruction modeling also has an indirect effect on implementation by way of content: participating teachers learn from the online materials and incorporate this new knowledge into their work with students. In this way, better content contributes to better implementation. More ambitiously, instruction modeling can help with the high end of implementations: by identifying the mathematical knowledge and skills teachers need in order to provide truly exceptional instruction as they work with individual students or small groups, instruction modeling can offer a template for advanced teacher professional development.

In the matter of *interface*, my own opinion is that instruction modeling is of relatively little help. After all, the whole premise of instruction modeling is to preserve the *content* even as the *interface* is radically altered, transformed from a traditional classroom to blended learning. The expert teachers whose knowledge is the source of the content have no special insight into what qualities the new interface should have. Fortunately, learning science offers abundant guidance in this matter. This is why a designer who employs instruction modeling should not stop at instruction modeling alone; learning science is an invaluable complement.

In conclusion, it is true that instruction modeling is a laborious approach. Moreover, it is not by itself a guarantee of success: blended learning designers must also rely on learning science and other sources of guidance to ensure that their programs are strong in each of the essential dimensions. Nevertheless, this complexity should not daunt us. Those inexperienced in education often imagine that there must be some simple solution, some easy way to provide outstanding instruction to all students. The reader who has made it thus far in the present book knows this to be false. There is nothing simple about good education, and no method can succeed without correctly identifying good instruction, getting the details right in the content of every lesson, and ensuring that classroom teachers are deeply knowledgeable in the art of instruction in their subjects. In other words, the challenges inherent in instruction modeling are present in some form in *any* approach that has a chance at success. The difference, however, is that instruction modeling offers us a guiding star: instead of creating everything from scratch, we can look to existing traditions of outstanding instruction, using their accumulated wisdom to craft excellent curricula, good lessons, and exceptional professional training for teachers.

How Can We Know If a Program Is Effective?

By now we have seen how instruction modeling *could* improve the quality of blended learning, and how this *might* lead to better education. But does it really work? In other words, when a program designed using instruction modeling is actually implemented, does it yield an improvement in instruction over the status quo?

Since Reasoning Mind has developed several programs—each relying heavily on instruction modeling, and collectively used by hundreds of thousands of students over the years—we can look to these programs as suitable tests of the question. However, before we can undertake this, we must first answer a broader question: how can we know if a blended learning program is effective?

In chapter 1, we saw that the naive way of assessing the quality of a program—viz., by looking at the test scores of participating students—is inadequate for the task. Test scores do provide some information, but only in the context of carefully designed scientific studies. Multiple studies are needed, and even then they offer only a partial picture, and thus must be supplemented by other evidence: the characteristics of the program itself, evidence about how it is used in practice, the reviews of students and teachers, and anything else of value that can be obtained. This difficult task of *program evaluation* is less like taking a patient's temperature and more like being asked to assess his general physical, moral, and intellectual health.[6]

Table 6.1 shows some of the categories of relevant evidence. This is not meant to be an exhaustive list, but it includes the forms of evidence I have most often seen collected.* The list includes three "design" categories and four "outcomes" categories; it is natural that outcomes by themselves are insufficient, since—as chapter 1 explained—there are many aspects of good instruction whose effects are not practicably measurable. The "design" evidence falls into the three groups introduced before: content, interface, implementation.

Many of the items in Table 6.1 are self-explanatory, but some bear commentary. Test outcomes include the results of large randomized studies: these experiments, as chapter 1 explained, are the gold standard for measurement. In a large randomized study, entire schools are normally randomly assigned to be in either the experimental group (which receives the program) or the control group (which does not). However, we also saw that even these studies were subject to many serious limitations. To obtain a better picture, we need to look at multiple field experiments: conducted with different student groups, using different outcome tests, and carried out at different points in time. Obtaining many large randomized studies is close to impossible, since they are exceptionally costly and difficult to do. For this reason, it is valuable to also use weaker designs whenever they are available.

* "Most often" is a generous phrase. Many evaluations look only at test scores, or supplement them only with a smattering of classroom observations. Of course, the best evaluators are more thorough. In my view, the categories in this table make up a minimal list of what should be considered if a reasonably complete picture is sought.

TABLE 6.1. Categories of Evidence Commonly Used in Program Evaluations

	Category of evidence	*Specific types of evidence*
Design	1. Quality of content	Quantitative: verification that content conforms with scientifically derived principles of good instruction. Humanistic: expert reviews; historical evidence (regarding the use of same or similar content).
	2. Quality of interface	Verification that interface conforms with scientifically derived principles of effective multimedia learning.
	3. Quality of implementation support	Verification that adequate teacher training and support are provided. Expert reviews of implementation approach.
Outcomes	4. Test outcomes	Field experiments (including, but not limited to, large randomized studies). Correlational evidence.
	5. Student, teacher, and administrator attitudes	Surveys. Interviews.
	6. Usage data	Time online.
	7. Classroom observations	Student time on task, engagement, boredom, frustration. Verification that actual implementation conforms with intent.

One such design is the *small, student-level randomized study*. As the name suggests, this design assigns students—rather than entire schools, as a *large* randomized study might do—randomly to different treatments. This study has the benefit of randomization, but usually has some problems with controlling for equivalence between the two conditions. For example, if both groups of students share the same teacher, how can we be sure that the teacher did not allow his own biases to interfere with his lessons, unintentionally showing more enthusiasm for one of the two conditions? Nevertheless, this design—if done properly—can be reasonably strong.

Another widely used design is the *quasi-experimental study*. In this design, the two groups are not randomly selected, but *are* chosen to be as similar as possible (for example, making sure that the students in the experimental and control groups are comparable in test scores, grades, demographics, and every other indicator we can find). The name "quasi-experimental" is unfortunate, since this form of study is beyond question an experiment: after the two groups are chosen, something is deliberately altered to create a difference between two conditions. What is lacking is *randomization*—and it is a material lack. We can be sure that our two groups are comparable in all of the attributes for which we have data, but we can never be certain that there is not some important but unmeasurable difference in students' personalities, school cultures, principals' leadership styles, parental attitudes, or some other such thing. Nevertheless, well-designed quasi-experiments are considered to provide good evidence. In particular, they often benefit from size: this design makes it easier to include more students, thus yielding more generalizable findings.

Finally, there are *correlational studies*: studies that are not experiments, but look at data after the fact to identify trends. For example, we can assess whether students who spent more time online in a blended learning program ended up disproportionately improving their test scores. While weaker than a true experiment, this design still offers *some* evidence when done properly. Many prominent education researchers work exclusively with correlational data and have been able to extract deep and important insights from it. One benefit of correlational data is that it can be massive: no experiment needs to be conducted, so thousands or even millions of records can be included. This means smaller effects can be statistically detected.

It also means findings are more likely to generalize to other groups of students.

In assessing a program from the perspective of test scores, all of these designs can be helpful. After marshaling all of the available studies, a judgment can be made from the totality of the evidence, appropriately discounting studies in proportion to their individual design flaws. Of course, some studies are flawed beyond redemption; but, as the previous examples illustrate, there is a wide range of study designs that are useful despite some imperfections.

Moving on, another type of evidence listed in Table 6.1 are surveys of teachers and students. Such surveys might ask students and teachers whether they like a certain program, whether they believe it was helpful, and how it compares to their prior experiences. Of course, people are biased. Students in particular often prefer worse courses to better ones[7]—as any college professor who has endured student course evaluations will bitterly confirm. However, while surveys cannot be taken at face value or used alone in evaluating a program, they do offer a helpful additional piece of information. For example, the extent to which students and teachers like a program is meaningful in itself. And if a large fraction of students or teachers report that a program is unpleasant, harmful, or irrelevant, then we would be rash to disregard the warning.

The final item from Table 6.1 that I would like to discuss is usage data, and time online in particular. A skeptic might ask: why is this relevant? After all, who cares how long students spend using a program? Are not results the only thing that matters? If anything, our skeptic would argue, a program that gives good results in a short span of time is *better* than one that accomplishes the same thing but requires longer. This argument misses the mark for a familiar reason: our skeptic is too sure of the precision of these "results" he mentions. In reality, our picture of any program's outcomes is an impressionistic one, and must be supplemented with other evidence. If a program is used by students for only one hour but shows a large effect on test scores, this means not that the program is wonderfully efficient, but rather that the measured effect is spurious.

How much time online is enough for a program to plausibly make a serious contribution to the overall quality of instruction? The more the better. But obviously a program cannot improve mediocre instruction to the point of excellence without taking up the *totality* of students' time devoted to the subject in question: after all, instruction cannot be excellent on the whole if it is excellent for part of the time and mediocre for the remainder. If a program is used for ten or twenty hours a year, what reasonable person would expect it to matter at all? If it is used for thirty or sixty, then it can make a meaningful contribution but never transform students' education. But if it is used for a hundred hours a year or more, then we are on to something significant. Learning takes time. There are no shortcuts.

This matter of time online is no straw man. The vast majority of blended learning programs are used lightly as diversions. Why waste time studying their effectiveness? Why attempt to measure the instructional contribution of a program that plays no role in instruction beyond a cameo appearance?

At this point, the attentive reader may raise a doubtful brow: did I not recount, in the introduction to this very book, how a five-week stint at a summer camp changed the course of my life, raising me from mathematical ignorance and setting me on the path of virtue? Indeed, this is true. But the five weeks were devoted entirely to mathematics: all told, I must have spent two or three hundred hours on the subject in this time, more than a high school student ordinarily spends in a full year. And the first summer at Mathcamp served mostly to secure my interest; I returned two more times, and also came to spend considerable time during the school year reading mathematical books, learning to solve more advanced problems, and the like. If I ended high school with a stronger mathematical preparation than my peers who had not attended Mathcamp, it was because I had spent at least one or two thousand additional hours—the equivalent of several years of added instruction—on learning the subject. Even a program like Mathcamp, which is as perfect an example of brilliant instruction as I have seen, does its work over long periods of time. My story is no counterexample to the rule that even the best programs require time; it is, rather, only one more proof.

The Empirical Evidence for the Effectiveness of Instruction Modeling

Having identified the forms of evidence that must be considered, we can now ask: what can be said about the effectiveness of the instruction modeling programs developed by Reasoning Mind over the years? We will go through the seven categories from Table 6.1 of the preceding section, referring as we go to the body of studies on these programs. These studies have been carried out by researchers from a wide range of institutions, including staff scientists at Reasoning Mind, researchers at SRI International, professors at Columbia University Teachers College, the University of Houston, and Texas A&M University, and in one case even an undergraduate student at Harvey Mudd College. I have divided the studies into three groups: peer-reviewed publications (Table 6.2a), independent evaluations (Table 6.2b), and other studies (Table 6.2c). The tables include Cohen's *d*, which is the number of standard deviations by which the treatment group outperformed the control group, wherever this statistic was reported or could be calculated from other reported statistics.

We begin with the *content*, the first item in Table 6.1. Of course, the quality of the content is the primary motivation for instruction modeling in the first place. In a program like *Edifice*, the process of modeling is so meticulous that the content cannot be anything but close to the source. In a program like *Foundations*, the fidelity of modeling is somewhat lower, but nevertheless captures much of the original content. This was the subject of Maia Valcarce's college senior thesis (Study #12 in Table 6.2c), which compared Foundations to the traditional practice of Russian mathematics education. The ability of the programs to accurately represent the content of the original source is also documented by Studies #2 and #3. All told, the evidence confirms that these programs have the intended high quality of content.

Moving on to the *interface*, we find a more mixed picture. Study #3—along with chapter 4 of the present book—documents the great extent to which Edifice conforms with the recommendations on learning science on this subject. The same can be said of Blueprint. Meanwhile, Genie I and Foundations are clearly in partial, but not full, conformance. For example, these programs do not separate sound from text, instead offering voice only in the form of narration of displayed printed material. Nevertheless, they do follow certain other recommendations, such as avoiding extraneous motion or sounds during instruction. All of the programs place a strong emphasis on encouraging teachers to work individually with students, which contributes positively to the richness of the interface (understood

TABLE 6.2A. Peer-Reviewed Studies of the Effectiveness of Instruction Modeling Programs Developed by Reasoning Mind

Study	*Type; Program*	*Findings*
1. Evaluating the Reasoning Mind fifth-grade common core curriculum[a]	Quasi-experiment; Foundations	Outperformance on Singapore math test ($d = 0.53$)
2. Reasoning Mind Genie 2: an intelligent tutoring system as a vehicle for international transfer of instructional methods in mathematics[b]	Descriptive; Genie I and Foundations	Positive student and teacher attitudes
3. Adapting international math curricula using an online platform[c]	Quasi-experiment; Edifice	Outperformance on Reasoning Mind mathematics content test ($d = 0.83$)
4. Field observations of engagement in Reasoning Mind[d]	Observational; Foundations	High time on task, high engagement, low boredom
5. Incorporating effective e-learning principles to improve student engagement in middle-school mathematics[e]	Observational, quasi-experiment; Edifice	Increased time on task, increased engagement, and lowered boredom in treatment group vs. comparison group

[a]Mingle et al., 2017b.
[b]Khachatryan et al., 2014.
[c]Mingle et al., 2017a.
[d]Ocumpaugh et al, 2013.
[e]Mulqueeny et al., 2015.

in the broad sense of encompassing *all* instructional interactions, not only ones with the computer). It should be noted that any shortcomings in the interface of some programs are not the result of the use of instruction modeling, and therefore should not be seen as an argument against the approach; in fact, to the contrary, those programs which model most meticulously (for example, Edifice) also

TABLE 6.2B. Independent Evaluations of the Effectiveness of Instruction Modeling Programs Developed by Reasoning Mind

Study	*Type; Program*	*Findings*
6. An evaluation of the Reasoning Mind pilot program at Hogg Middle School[a]	Small randomized study; Genie I	Outperformance on Reasoning Mind mathematics content test (***d* = 1.66**), Texas TAKS test (***d* = 0.79**), and Stanford Achievement test (***d* = 0.48**); positive student attitudes
7. An evaluation of the 2005–2006 Reasoning Mind project[b]	Small randomized study; Foundations	No statistically significant improvement on test scores (***d* = 0.07**), likely due to failed implementation
8. An evaluation of the 2006–2007 Reasoning Mind program[c]	Quasi-experiment; Foundations	Outperformance on Reasoning Mind mathematics content test (***d* = 0.80**, ***d* = 0.53**, and ***d* = 0.43** at the three participating schools); no effect on Texas TAKS test; positive student and teacher attitudes
9. Evaluation of the 2010–2011 Reasoning Mind program in Beaumont ISD[d]	Quasi-experiment; Foundations	Outperformance on Texas TAKS test (***d* = 0.16**); positive student and teacher attitudes
10. Independent evaluation of Reasoning Mind programming[e]	Correlational study; structured interviews and observations	Improvements from pre- to post-test on test scores (program test, Texas STAAR), positive student and teacher attitudes
11. An Efficacy Study of a Digital Core Curriculum for Grade 5 Mathematics[f]	Large randomized study; Foundations	No effect on West Virginia WVGSA test

[a]Weber, 2003.
[b]Weber, 2006.
[c]Waxman and Houston, 2008.
[d]Waxman and Houston, 2012.
[e]Boriack et al., 2015.
[f]Shechtman et al., 2019.

TABLE 6.2C. Other studies of the Effectiveness of Instruction Modeling Programs Developed by Reasoning Mind

Study	*Type; Program*	*Findings*
12. Russian mathematical pedagogy in Reasoning Mind[a]	Descriptive; Foundations	Foundations faithfully captures much, but not all, of Russian mathematics education practices
13. Report on the 2011–2012 Reasoning Mind Implementation at Dallas ISD[b]	Correlational study; Foundations	More time online corresponds to more test score growth (STAAR, ITBS); positive student, teacher, and principal attitudes
14. Designing for consistent implementation of a 5th grade digital math curriculum[c]	Descriptive; *Foundations*	Strong implementation practices
15. Strength of research for Reasoning Mind[d]	Review of other studies	Evidence as a whole indicates effectiveness

[a]Valcarce, 2012.
[b]Douglas et al., 2012.
[c]Roschelle et al., 2016.
[d]Roschelle et al., 2015.

conform to the recommendations of learning science to the greatest extent.

As to the *implementation*, the final item in the group of three "design" categories, there is considerable evidence that Reasoning Mind's programs are quite strong. The extensive teacher professional development and support are described in Study #2 from Table 6.2a. This is also the subject of Study #4, which documents the intensity of Reasoning Mind's implementation work. While this is certainly a merit of Reasoning Mind's programs, it is unclear to what extent it can be attributed to the organization's use of instruction modeling. As I argued earlier, instruction modeling contributes chiefly to the quality of content, and plays a mostly supporting (albeit not negligible) role in the quality of implementation.

Moving to the "outcome" categories, we begin with *test scores*. A wide variety of studies have been conducted, and most have found a positive effect on test scores (with three cases of a neutral effect and no cases of a negative effect). The studies have used a wide range of different tests: the Iowa Test of Basic Skills (ITBS), the Stanford Achievement Test, the Texas Assessment of Knowledge and Skills (TAKS), the State of Texas Assessments of Academic Readiness (STAAR), the West Virginia WVGSA, the Singapore Mathematics test, and curriculum content tests developed by Reasoning Mind. There has also been a wide variety of contexts (different states, grade levels, student demographics, and school types). There has also been a wide range of study designs: randomized experiments, quasi-experiments, and correlational studies.

All told, these studies suggest that programs developed using instruction modeling—or at any rate, these particular programs developed using instruction modeling—have a positive effect on test scores, or in the worst cases a neutral effect. However, as I have already warned, we should not give tests a monopoly on our educational judgements; this evidence is encouraging for instruction modeling, but should only be considered in concert with the rest.

There is an interesting observation to be made in reference to these test outcomes. If one looks closer at the size of the effects in the various studies, a pattern emerges. The effects are largest when Reasoning Mind's own curriculum test is used, and smallest (though still positive much of the time) when high-stakes tests are used. For independent tests—such as the Singapore Math test—they are in the middle.

To put this in specific terms, let us consider the studies with reported values of Cohen's *d*. Out of five results reported on Reasoning Mind content tests, all five are positive with very substantial effects (Cohen's *d* of 0.83, 1.66, 0.80, 0.53, and 0.43). The two results on independent tests (Singapore Math test and Stanford Achievement Test) are also both positive (Cohen's *d* of 0.53 and 0.48), with effect sizes towards the low end of the range for Reasoning Mind content tests. For state high-stakes tests, there are two positive effects (Cohen's *d* of 0.79 and 0.16), two null effects (Cohen's *d* of 0.07 and –0.06, neither large enough to be statistically significant), and one study which did not report Cohen's d but found no statistically significant effect (Study #8). You should draw your own conclusions, but to me this indicates that Reasoning Mind programs certainly improve student knowledge of topics prioritized in the Reasoning Mind curriculum, and most likely also improve student knowledge of mathematics in general; however, the evidence does *not* suggest a substantial effect on same-year high-stakes test scores.

This trend is exactly what we would expect from the *nefarious alignment effect,* whose disagreeable acquaintance we made in chapter 1. After all, Reasoning Mind's own tests will always have the highest degree of alignment with the program. High-stakes tests are tightly aligned to the instruction of control groups, since the status quo in American schools is an intense focus on the material covered on such tests. Independent tests have a looser alignment with both Reasoning Mind and control group instruction.

The various *attitude surveys* of students, teachers, and administrators have been very positive. Studies #6, #7, #8, #2, and #10 all collected this variety of data, and generally found approval ratings among teachers and students of 90 percent or more. When asked what they dislike about the program, teachers often mention the challenges of technology (ensuring enough working computers, internet connectivity, etc.), the difficulties of scheduling (including reaching the recommended time online), and the alignment (i.e., that the programs spend time on material which does not appear on end-of-year high-stakes tests). Not surprisingly, these are exactly the issues chapter 5 discussed as barriers to dissemination of comprehensive programs. When asked what they most like about the program, teachers often cite the quality of the content, the extent to which the program engages students, and the strength of the implementation support. All told, this evidence suggests that programs relying on instruction modeling can be palatable to—and even the subject of some enthusiasm among—students, teachers, and administrators.

Interestingly, there is some evidence (Study #3) that students enjoy Edifice less than Foundations. Cited reasons include the timers attached to all of the questions in Edifice, and the less game-like nature of the program. These differences make Edifice a more faithful model of actual classroom instruction. Thus, more accurate modeling does not necessarily increase students' enjoyment of a program. But perhaps this is obvious: no student has ever written in to request better modeling.

As for *usage*, Reasoning Mind's records of time online show that when the programs are used as core curricula, students do indeed spend well over a hundred hours on the online lessons. When Foundations is used as an "ancillary" program (meaning that it is not the core curriculum, but still more extensively used than a "supplementary" program would be), students general spend in the range of thirty to sixty hours online. While less than the recommended time of seventy hours, this is nevertheless a considerable degree of exposure. This evidence indicates that between the implementation support and the inherent value which teachers find in the programs, the programs are in the end used enough to matter.

Classroom observations are our final category of evidence. Here, the evidence confirms that implementations are generally as intended: students work online, and teachers spend their time both facilitating the class and working with students individually and in small groups. One of the most interesting types of observation is found in Studies #4 and #5; these studies used a method called the Baker-Rodrigo-Ocumpaugh Method Protocol (BROMP), developed by Ryan Baker and his colleagues. BROMP allows trained observers to reliably measure students' behavior (including percentage of time spent on task) and affect (boredom, confusion, concentration, frustration, and so on). Time on task is obviously an important indicator, since students do not learn much when they are off task—or at least not much of what they are *supposed* to be learning. Affect is important, too: for example, studies have found that *concentration* bodes well and *boredom* bodes poorly for students' future academic success. Study #4 found Foundations students to be on task 89 percent of the time, a rate far higher than the typical rate of students in conventional classrooms (which is around 70 percent). Study #5 found even higher time on task for Edifice: students were on task 93 percent of the time, compared to only 69 percent for the control group. To put these findings into context, this increase of time on task is comparable to adding an additional semester of instruction. The studies also found considerable increases in concentration and reductions in boredom.

To what extent is instruction modeling responsible for these effects? Is not *any* form of blended learning inherently more engaging for students than ordinary instruction? Actually, no. Blended learning is no guarantee of heightened student engagement.[8] While it is possible that some *other* aspect of Reasoning Mind's programs is responsible for the high student engagement, it seems likely that instruction modeling would play at least some role. After all, the high quality of the content is attributable to instruction modeling, and it seems plausible that better content would be more engaging—just as it is in ordinary classrooms. And the fact that engagement is even higher in the program which models more faithfully is a further reason to suspect a role for instruction modeling.

In conclusion, the weight of the evidence points to the value of instruction modeling in developing effective blended learning. The programs Reasoning Mind developed have successfully improved the quality of mathematics instruction offered to hundreds of thousands of students. In my view, this offers a compelling case for the value of instruction modeling in spite of the challenges. I hope that more of those who develop blended learning programs—and everyone else involved in the education of children—will come to understand both the pivotal importance of high-quality content, and the value of instruction modeling as a way to consistently produce and deliver it.

Chapter 7

A Comparison with Other Approaches

INSTRUCTION MODELING IS ONE PARTICULAR APPROACH that can be used in designing blended learning programs. As we have seen already, it is not meant to be used in isolation: good designers combine it with such approaches as usability testing, iterative design, and the application of learning science.

In this chapter, we will cover the relationship of instruction modeling to the most popular approaches in blended learning. We will not discuss approaches that are universally used in any good software design—such as, for example, usability testing—instead confining the discussion to those particular to blended learning.

To what extent is *any* approach valuable? In the preceding chapter, we introduced three dimensions of quality for blended learning—content, interface, and implementation—and argued that a blended learning program offers excellent instruction if and only if it is strong in each of these dimensions. Accordingly, we can assess the value of an approach against its contribution to one or more of the dimensions.*

* Incidentally, this line of reasoning illustrates the importance of using at least *some* approach. If a designer sets about building a program with no well-substantiated plan for ensuring its quality, the result will be predictable.

Instruction Modeling. George A. Khachatryan, Oxford University Press (2020)

DOI: 10.1093/oso/9780190910709.001.0001

For example, the last chapter made the case that the special value of instruction modeling was in ensuring the high quality of the *content* in blended learning.

In reviewing each approach, we will ask two questions. First, does the approach make instruction modeling redundant? If, for example, there are approaches that ensure outstanding content with no recourse to instruction modeling, then we would have no need for it. The reader will not be surprised to find that I do not believe there are such approaches—at least not among what is currently widely practiced. In this chapter, I will explain why this is the case.

Our second question concerns whether a given approach can be combined with instruction modeling to obtain the benefits of each. Most of the time, the answer is in the affirmative. We will discuss the interplay between instruction modeling and each approach, which—in my view—is often quite interesting.

Featurism, the Monster with Only One Head

Before we discuss more substantive approaches, we must address the weed of the garden of blended learning: *featurism*. You may have never heard the word, but you have surely encountered the concept. Featurism is the latching on to some one feature of instruction, claiming that this feature by itself will carry the day and ensure instructional success. It is a weed that peddles itself as a simple educational panacea. Like many weeds, it comes in different varieties:

"This program will work because it is all about games. Children love games! Online games for everyone. It can't fail!" Featurism *ludensis*.

"This program is special because it adjusts to the learning styles of students. Visual learners get animations, auditory learners get voice descriptions, and kinesthetic learners get jigs. With this degree of personalization, everyone can't help but learn perfectly!" Featurism *stylensis*.

"This program is all about online videos. The videos explain things like a teacher, but now students don't feel the pressure of a normal class. They can pause, rewind, and repeat as needed. It's much better!" Featurism *onlinevideënsis*.

There are a thousand more. Each is unique, but they share in the same essence: the belief that good instruction might plausibly come from some single clever idea.* If you ever hear, "This program works because of _____," you are most likely speaking with a practicing featurist.

Is featurism bad? How do we know there *isn't* some one idea that unlocks the doors of unlimited learning? Why discard an approach simply because it proposes such a magical key? To answer this objection, we don our common sense caps.** Is good education *ever* the product of some one magical idea or practice? Search your experience. Recall your best teachers. What one magical practice made their classes so good?

My favorite teacher in high school was Dr. James Houlihan, who taught Latin. Hopefully he will not read this chapter, with its cavalier abuse of the language he so loves. Dr. Houlihan was deeply knowledgeable in his discipline, and could answer any question. Students joked that he knew everything about everything, but it was not entirely a joke. He chose wonderful readings for his classes. He was a masterful moderator, giving free reign to interesting discussions but also knowing when to gently curb digressions. His love of ancient literature was obvious and infectious. He did not stop at grammar, but introduced us to every work in its proper historical, philosophical, and aesthetic context. There is no *one* thing which Dr. Houlihan did to teach Latin literature well. He did a vast array of things well, in coordination, every day.

If your experiences are anything like mine, you will recall the same about your favorite teachers. They were not featurists. And if we seek more respectable sources of evidence than our own memories, we can look to ethnographies of education. Studies of excellent teaching—whether Liping Ma's excellent work[1] or the Trends in International Mathematics and Science Studies (TIMSS) video studies[2]—show that good teaching (and good curriculum development, for that matter!) is complex and comprehensive. Studies of

* Of course, sometimes featurism will combine two, three, or four ideas. The upshot is the same: a focus on cherry-picked details instead of on the whole picture of what constitutes good instruction.

** An item of apparel often neglected in discussions of blended learning. We will not make the mistake.

the requisite knowledge for good instruction, such as Deborah Ball's famous work on the subject,[3] show that it is vast and broad; there is no "thing good teachers know": there are a *thousand* such things.* Perhaps we cannot *prove* that no featurist approach can ever work, in the same way in which we cannot prove that the Loch Ness Monster does not exist. But until someone produces a more convincing photograph, we should go with our common sense.

The opposite of "featurist" is "comprehensive." This is why instruction modeling, for example, is not featurist: the entire premise of instruction modeling is to capture *all* of the facets of excellent instruction as found in existing classrooms. It is valuable precisely because it offers a practical way to attain this completeness. It is an antidote to featurism. Of course, even then a good designer must employ additional methods beyond instruction modeling to look after the quality of the *interface* and the *implementation*, as we discussed in the preceding chapter. A good instructional designer cannot be a one-trick pony. He cannot even be a two-, three-, or four-trick pony. A good instructional designer is a pony of a thousand tricks.

It did not take much to reveal the folly of featurism. Why, then, is it so popular? And popular it is: I would say that the substantial majority of today's programs are at least to some extent featurist. I believe that some role is played by commercial pressures. Designing a program that is truly comprehensive is a great undertaking, and not one that is financially rewarded: as we saw earlier, more comprehensive programs are actually *more* difficult to disseminate. For-profit developers do the natural thing in avoiding the fiscally counterproductive effort.

Another draw of featurism is that it makes for simple, memorable slogans. A marketer would say that it allows for a crisp market positioning. Everyone will remember, for example, "the program that works because everything is taught through games"—it is a simple premise that sets the program apart from others. Truly comprehensive programs, on the other hand, face the uphill battle of explaining

* This is true beyond education. If a violinist told you that her special virtuosity came from the unique angle she used to position her bow, you would judge it to be a joke. As far as I am aware, there is no reasonably complex realm of human endeavor in which excellence is the product of finding "the right trick."

their mass of complexities—a rotten positioning indeed, if it can even be called that.

How can you spot featurism, telling its manifestations apart from the productive plants in the technological garden? I would propose two litmus tests. The first test is something you could call "conventionalization." The test is done in the mind: simply imagine the program transformed into conventional instruction. For example, if a program is entirely made up of games, imagine a teacher who instructs students exclusively by playing analogous games in person. If a program consists entirely of videos that can be rewound and paused at will, imagine a teacher who works exclusively by lecturing, with students able to pause her or ask her to repeat something. If the conventionalized version seems silly and flat, then the program has tested positive for featurism. After all, if a set of instructional experiences is inadequate when delivered offline, why would it suddenly become adequate when moved online?

The second litmus test is more prosaic: simply evaluate the program against the three dimensions of quality introduced in the last chapter. Has the program convincingly produced good *content*, using either instruction modeling or some other method to ensure consistently high quality? Does it have an *interface* that conforms—at least to a reasonable extent—with the recommendations of learning science? Is there robust teacher training and support in place to ensure strong *implementation*?

Featurism is rampant. The remedy is education—this time, not of students, but of educators. The more people understand that there is no substitute for strong, comprehensive instructional design, the easier it will be to make such design a reality. Then, at last, the garden of blended learning will be free to yield the rich harvest it has promised for so long.

Applied Learning Science

One particularly important approach is *applied learning science*. In their book on this subject, Hess and Saxberg refer to the approach as *learning engineering*,[4] since—like engineering in other disciplines—it consists of the systematic practical application of experimentally derived principles. As we saw in the first chapter, the last several decades have seen a profusion of excellent research in education psychology, and it would be a great waste indeed to ignore this source of guidance.

Does applied learning science make instruction modeling redundant? Naively, we might think that it does. After all, learning science offers judgments on all manner of educational questions; as learning science progresses, would we not eventually end up with a complete understanding of what makes for good education? And armed with this understanding, could we not create perfect blended learning programs? The attentive reader of the first two chapters will not fall for this trap. I will not repeat here all of the points made there, but the upshot is that in education, experimental research is a potent, but not omnipotent, tool: it can identify many (though not all) of the principles of good instruction, but applying these principles in each minute of instruction remains a difficult task that requires other sources of guidance. For example, knowledge of the laws of harmony is necessary, but not sufficient to compose good music. Knowledge of the psychological laws of perception and attention is helpful, but not sufficient to design good websites. It is the same in education.

Next we ask the reverse question: does instruction modeling make applied learning science redundant? After all, instruction modeling is a systematic way to capture and replicate excellent instruction. Why does it need scientifically derived principles, when it obtains its material directly from a battle-tested, certifiably outstanding source? This line of reasoning, too, is flawed. To begin with, learning science is helpful in identifying good traditions of instruction to model in the first place. An instruction modeler who does not know learning science might nevertheless choose well, but undoubtedly will do so with a higher probability if she knows learning science.

But beyond this, we saw in the preceding chapter that instruction modeling gives comprehensive guidance for only one of the three dimensions of quality for blended learning: *content*. Applied learning science helps tremendously with another, namely *interface*. For instance, instruction modeling can tell us that a situation calls for providing a certain explanation, but will not help us to know whether it is acceptable to use text, or if sound is better; learning science can provide invaluable assistance with such questions.

Even in the matter of *content*, learning science can be of use to the instruction modeler. For example, I often found knowledge of education psychology helpful when interviewing expert teachers and deciding on how to model their practices. A teacher might say: "I want students to do their homework at home, not at school, since I find that they remember the material better this way." Knowing that memory research has corroborated this teacher's view (we are more likely to remember something when we have encountered it in different environments[5]), I am likely to go to greater lengths to faithfully model her practice, even if it is difficult. If instead learning science had contradicted the teacher's beliefs, I might have been inclined to compromise on this point, choosing expediency over fidelity of modeling. This is just one example, but there are many other ways in which a knowledge of learning science helps to better understand and model expert teaching practices.

Finally, learning science often expands the possibilities for blended learning, broadening the instruction modeler's toolkit. For example, education psychologists have found ways to use "natural language processing"—the ability of computers to make sense of freeform text typed by a human being—in educational programs.* An instruction modeler could harness this in designing lessons: instead of asking only closed-ended or multiple-choice questions, she could design lessons that ask students to type answers to questions like "What would you do next in this problem?" The computer could make (some degree of) sense of the student's answer and respond appropriately. This would yield a more accurate model, since a teacher often asks such open-ended questions in class. Thus, the work of learning scientists can give instruction modelers the tools to build more faithful models.

These points illustrate the close interplay that can exist between instruction modeling and applied learning science. Learning science not only complements instruction modeling (e.g., by informing decisions concerning the interface of lessons), but also directly supports it by strengthening the quality of modeling.

Are all applications of learning science good? Certainly not. First, one often encounters "learning pseudoscience": programs that are ostensibly designed following principles of learning science, but in fact are based on nonsense. This includes, for example, programs

* We take up this example in more detail in the final chapter, which discusses the future of instruction modeling. Natural language processing has enormous potential in instruction modeling.

that adjust for different "learning styles,"[6] a notion that has been repeatedly debunked by research. We will return to this example in the next section. But beyond learning pseudoscience, even correct learning science can be misapplied. The most common form of misapplication is one we have already met: *featurism*. Yes: learning science, too, is susceptible to featurism.

This happens when a program is designed based on some isolated ideas from learning science, with no comprehensive attendance to *all* of the aspects of good instruction. Take, for example, the finding from memory research I mentioned just now, viz., that things are better remembered when they are encountered in a variety of different environments. If you and I were featurists, we could design a program around this finding: a mobile phone application that uses the phone's built-in location service, sending mathematical problems whenever the student changes location. We would think of a catchy name: "Motion Math Mate," or "Math-on-the-Go." We would make a website touting the firm scientific grounding of our application. It would include fancy graphs and citations of scientific studies. We could even obtain an endorsement from an actual memory researcher! Our brochures would be convincing indeed: "Have you ever wondered why your students don't remember more of what you teach? Today's schools teach everything in the same location, the classroom. But the latest learning science shows how inefficient this is—we remember far more when we are exposed to information in a variety of environments. The school, the home, the shopping mall, the amusement park. With Math-on-the-Go, you can harness the incredible power of cutting-edge learning science for the benefit of YOUR students! Join the revolution, for only $19.95."

Of course, the program itself would be terrible. The content would be weak—a collection of routine problems, no better than those found in mainstream textbooks. The interface would be whatever we could cheaply program and fit on a phone screen. There would be no teacher training or implementation support—after all, who can afford it for $19.95? But purchasers would not realize it, and—to be fair to the developers of featurist programs—neither would we. We would all think that we were bringing the benefits of a valuable finding of learning science to students. We would be very pleased with ourselves.

Hess and Saxberg did well to describe applied learning science as *learning engineering*. Like all engineering, done well it is a comprehensive discipline. If a structural engineering firm "specialized in Newton's second law of motion" and used this to "build bridges of exceptional sturdiness," you would be willing to drive a good distance to avoid traversing their products. A proper engineer should know and apply a *comprehensive* set of principles, drawing from mechanics, materials science, wave motion, and who knows what else. Learning engineering is no different.

In conclusion, applied learning science is an invaluable tool. However, it does not answer all instructional questions, and in particular is of limited value in developing excellent *content*. This makes it an excellent pairing with instruction modeling. Even then, learning science must be applied comprehensively: cherry-picking individual principles is no way to do it. A proper "learning engineer" should possess broad knowledge of learning science and employ its full range to inform design decisions.

"Individualization," A.K.A. "Personalization"

Perhaps the most common argument advanced in favor of blended learning is that it "personalizes" education. This is a rebranding of what was once (and is still often today) called "individualization." The premise is that conventional instruction, with its whole-class "factory model," misses the individual needs of students. By adapting to each child, a computer can personalize instruction, improving the quality of education. Very many blended learning programs are built—and marketed—primarily on this idea.

There are several reasons why this idea has captured the public imagination. One is a general belief in the inescapable tedium of ordinary schooling. This has been much helped along by the writers of screenplays and novels, such as Mark Twain, who—when not busy eviscerating school boards—gave us such quotes as "Education consists mainly in what we have unlearned."[7] The image of row after row of bored schoolchildren has been forever stamped on our memories; "personalization" promises to free them from their bondage. A second reason is the idea's intuitive appeal: we expect customization from our barber, dentist, real estate agent, and so forth, and naturally think that more of it could only improve education. A third reason is a famous study published in 1984 by Benjamin Bloom; Bloom's study reported that personal tutoring yielded improvements of two standard deviations* in learning over ordinary whole-class instruction.[8] More recent work has cast doubt on the size of this effect,[9] and has also called into question whether personalization is indeed the mechanism responsible for the success of tutoring as an instructional method. Regardless, the study has done much to promote personalization as a universal solution—though, ironically, this had never been Bloom's intent.

Although the word "personalization" is often bandied about as if its meaning were obvious, the notion is in reality far from a monolith. Even taking it to mean "adjustment to differences between individual students," there are at least three** distinct strands of "personalization," each found in many different blended learning programs.

The first strand is *feedback*. In a whole-group lesson, the teacher typically asks a question and calls on one student to answer; this student then receives the teacher's feedback on her answer, but the other students do not. By contrast, thirty children working on online lessons can all receive immediate feedback on their answers to each of the lesson's questions. In this sense, instruction is "personalized" for each child. There is a good deal of research on the benefits of feedback.[10]

The second strand is *learning styles*. In its most common form, the theory of learning styles posits that different people learn better in different ways: "visual" learners do best with text or images, "auditory" learners with sounds, "kinesthetic" learners with movement, and so forth. Some programs claim to adjust to children's personal styles. Unfortunately, the theory of learning styles has been consistently debunked by experiments.[11] Many variations have been proposed, and all have met the same fate at the hands of the unforgiving laboratory. It appears that we are far more alike than we are different in how we learn: raindrops rather than snowflakes.

* For those less familiar with statistics, this is an enormous effect. It would move a student from the 50th percentile to the 98th.

** A fourth strand would be self-pacing. I will omit it here, as it was discussed in detail in chapter 5.

Despite this, there are *some* reasons to believe that different students might sometimes benefit from different treatments of the same topic. For example, Southern Methodist University's Candace Walkington has done experiments showing that children learn better from word problems whose scenarios match their interests.[12] Thus, a child who likes cupcakes but is indifferent to dragons would learn better when given a cupcake problem than when given a mathematically identical dragon problem.

The third strand is *knowledge tracing* (also called "student modeling" in the research literature). The idea here is to track what topics or concepts students already know, and then tailor instruction based on this information. This is helpful for two reasons: first, it improves efficiency by avoiding wasted time on things students already know; and second, it makes sure students are not asked to attempt to learn a concept when they are missing an important prerequisite. Thus, a program that uses knowledge tracing will prevent a student who already knows how to add whole numbers from being pointlessly retaught this topic, and also will prevent a student who does *not* know how to add whole numbers from being flung into a lesson on the addition of fractions. There is considerable research evidence for the benefits of knowledge tracing, and several well-known blended learning programs (such as the programs of Carnegie Learning and ALEKS) do some form of it.[13]

All told, "personalization"—especially feedback and knowledge tracing—is clearly a good thing. So does it make instruction modeling irrelevant? In other words, if we can build blended learning programs that personalize instruction, have we any need for the added difficulties of employing instruction modeling?

The answer is the same as before. Personalization, while clearly helpful, is no panacea. Personalization helps improve the quality of instruction, but excellent instruction is good in a great variety of ways, not just in its ability to adapt to individual needs. If we believe that personalization will by itself cure ignorance, we are succumbing to featurism.

A thought experiment* helps elucidate the issue. Imagine two mathematics classrooms. You must choose one for your child. The first classroom is a conventional classroom, but taught by an

* This is an example of "conventionalization," the litmus test suggested earlier.

exceptionally qualified teacher: everything is taught clearly, sequentially, logically, and in general conformance with all of the principles of excellent mathematics instruction. However, the lessons are planned in advance and are the same for everyone in the class. The second classroom is run by a teacher who magically knows exactly what concepts your child has understood, and focuses exclusively on concepts she has not yet mastered: in other words, it is a classroom of perfect knowledge tracing. However, aside from this the teacher is a disaster: he knows little of mathematics and even less of how to teach it. Which class would you prefer? I believe most of us would choose the first class: it may not be perfectly efficient, but your child would end her schooling with a solid grasp of the subject. Personalization is simply no replacement for good instructional content. If we personalize bad instruction, we obtain bad instruction that adjusts to be bad in a different way for each child. It is a suit custom-tailored by the best of Saville Row, but with Styrofoam in place of fabric.

The clever reader, ever seeking a loophole, might suggest a third classroom: one which is *both* personalized *and* instructionally strong. And of course, this would be the best of all possible worlds. This is why instructional designers who use instruction modeling should *also* attend to personalization. There are many ways this can be done; the most obvious concerns *feedback*: simply making sure that no opportunities to provide it are missed.

Knowledge tracing presents more of a challenge. As a matter of practicality, the world's great traditions of mathematics instruction—whether Russian, Japanese, or others—are built around whole-group instruction. (After all, no nation has been wealthy enough to educate all of its children through private tutoring.) Accordingly, it is possible to obtain very detailed descriptions of how expert mathematics teachers plan whole-group lessons, but the same information is not available in the case of tutoring. There *are* expert mathematics tutors, but they generally work with students who *also* attend regular mathematics classes. The question, "How would you educate students if you were tutoring them, and this would be their only source of mathematics instruction?" is unavoidably a speculative one. This illustrates one of the biggest limitations of instruction modeling: the strength that it is firmly rooted in proven practices carries the attendant weakness that it is bound to the existing, unsuited to excursions into the unknown.

One solution to this paradox is to do instruction modeling *within* topics and knowledge tracing *across* them. Foundations adopts this approach: students go through a sequence of "objectives," each covering a specific mathematical topic and written with reliance on instruction modeling. Meanwhile, the program automatically diagnoses past "objectives" that the student does not know and sends the student back to fill in these gaps.

Another solution is to mimic the conventional relationship between classroom teaching and tutoring. In this case, the program would include a sequence of linear lessons developed using instruction modeling, and a *separate* mode that would do knowledge tracing to remedy gaps in the student's understanding.* Just as conventional private tutoring usually plays the role of remedying gaps left from the regular mathematics class, the "tutoring" mode would identify and remedy gaps in the student's understanding of the concepts from the main sequence of online lessons. This division of labor is especially appealing because remedial tutoring usually focuses on the basics (i.e., the core ideas which are essential for future learning), whereas good instruction as a whole covers not only the basics but also a great deal of enrichment material—i.e., content that is not strictly essential by itself but serves to develop students' overall mathematical maturity.

As this example shows, sometimes combining instruction modeling with other valuable approaches—in this case, personalization through knowledge tracing—is not straightforward. Compromises may be needed in order to best enjoy the benefits of each.

À la Carte

A very common approach to blended learning is something that might be described as "à la carte." This means offering to teachers a wide range of different online programs, allowing them to assign bits and pieces of each to students. Schools that prefer "à la carte" often also mandate that participating programs be "interoperable," which in practice means that for every state educational standard (e.g., "students know how to add decimals to the third decimal place")

* At the time of this writing, something like this is under development for the Blueprint program.

there should be exactly one unit in each program. Then, teachers select the program they would like to use for each state standard separately, going through them in some order.

Two arguments are most often put forward for "à la carte." First, there is an appeal to "learning styles," since teachers are supposed to assign visual programs to "visual learners," auditory programs to "auditory learners," and something I cannot bring myself to imagine to "kinesthetic learners." And second, the approach allows schools to select their favorite programs for each topic; teachers or administrators might prefer one program's treatment of decimals, for instance, but another program's approach to common fractions. "À la carte" is said to let all flowers bloom (if you will forgive me for mixing horticulture with gastronomy).

We know now that there is not much to "learning styles," so this argument is not convincing. The second argument, however, appears plausible—at least at first glance. After all, is it not natural that different developers would succeed to varying extents with different topics? And if we accept this, then why would we not want children to have the best of each?

The problem with this line of reasoning is that it assumes that a mathematics course is a disconnected collection of topics, with no great need for cohesion. The thinker imagines, for example, a great art collection: it would have the best painting of Raphael, the best of Leonardo, the best of Ingres . . . and in this case, no one could deny that the collection as a whole was great, since it consisted of so many great works. However, as we discussed in chapter 2, mathematics is not like a collection of disparate works: the topics are closely intertwined, with each building on the other. If you change the treatment of one topic, then you must necessarily change the treatment of others. Thus, the best analogy would be not to a *collection* of paintings, but to a *single* painting, with the various mathematical topics corresponding to elements of the scene. The follower of "à la carte" would compose his painting by drawing each part from a different artist. He would argue that Raphael has the best cherubs, Leonardo is a master at depicting the ermine, and Ingres is without equal in the matter of necklines. His composition would have an odalisque by Ingres holding a Leonardo ermine, watched from above by a gracefully floating cherub of Raphael. You would not want it in your living room.

There is also a different, albeit related, problem with the "à la carte" approach: by forcing programs to have one unit for every state standard, "à la carte" imposes an unnatural structure. State standards are not written to be a table of contents for instruction; they are written only to lay down what students should know by the end of the year. Even though a good mathematics course does go through ideas in a sequential order, this is not tantamount to teaching every unit in isolation. For example, a course that covers addition of decimals and then proceeds to pentagons might wish to connect the two, giving problems that ask the student to find the perimeter of a pentagon with decimal side lengths; if each unit must be self-contained, then such problems have no place in the program. But these connections between notions are—as we discussed in chapter 2—essential for good mathematics instruction, and good courses brim with them.

For these reasons, "à la carte" is fundamentally incompatible with excellent mathematics instruction. A good course is a logical, carefully selected progression of mathematical ideas; each is treated in a way that hangs together with the rest;* and connections among the ideas are highlighted whenever possible, bridging concepts from arithmetic, algebra, and geometry.

Is there nothing then of value in "à la carte"? Not quite. First, the approach can be perfectly sensible if the goal is only to practice skills in preparation for an upcoming state test. In this case, a program is not charged with providing a complete mathematics education: it needs only to brush up on a variety of topics. And second, giving teachers some ability to assign practice on individual topics can help them more flexibly meet students' needs. For example, even a good program will not succeed with every topic for every student; teachers can benefit a great deal from being able to manually assign certain topics to students who need additional help.

Thus, "assignability" is an aspect of "à la carte" that *does* have value, and an instruction modeler should keep this in mind. Even

* Note that there are often many different ways to do this well. For example, one third-grade course might consistently teach word problems using bar diagrams (as in the Singaporean method), and another might use tables. Both are excellent ways, and they each work well when one is chosen and used consistently. But mix them into a hodgepodge, and children will be confused.

then, however, the best results are to be had when the "assignable" lessons are written in a way that coheres with the main course.

Big Data

"Big data" is in fashion these days: in academia, business, and yes—even in education. But what is it?

The field of statistics has been around for about a hundred years, with much of the classical work done in the first half of the twentieth century.[14] One of the main problems statistics has aimed to solve has been *prediction*. For example, imagine you were a marketer working for a hair products firm. Imagine furthermore that you had data on a thousand arbitrary households in the country—their income, number of children, preference for dogs or cats, and also their propensity to buy your hair gel. Then, you are given the same data—minus the data on hair gel—for a new, thousand-first household. Can you *predict* their likelihood of buying your hair gel? If you can, it will help your firm decide which new cities to enter, which stores to distribute through, and even how to advertise your product to appeal to the most likely buyers. Of course, predictive statistics is useful far beyond marketing: for entertainment, politics, engineering, and much else. Online video services would like to predict which films you will like (so that they can please you with their helpful recommendations), politicians would like to predict which districts they will carry (so that they can focus their energies on the true battlegrounds), and engineers would like to predict whether a car will crash if it continues driving (so that they can build cars that do not).

To do all these things, statisticians perform calculations on the data to produce formulas that make new predictions. This is called building *models*. In the 1940s, if you wanted to build such a model, you would do all of the calculations by hand: with a pencil, paper, and perhaps a slide rule. This imposed substantial limitations on the sorts of models you could build. They could not use too much data, or require too many computations. Otherwise, you would be stuck doing calculations for the rest of the year, decade, or lifetime.

In late twentieth century, all of this changed.[15] As computers became more powerful and more widely available, a vast world of more sophisticated models became possible. Of course, before any concrete models could be built, the foundational techniques needed to be developed (just as before we could all solve quadratic equations, someone first needed to discover the quadratic formula). Researchers began frantically developing these techniques. The researchers hailed from different fields: some were computer scientists working on algorithms, others were statisticians, and still others specialized in databases. Each group of researchers gave the new statistics a different name: "big data," "machine learning," "predictive analytics," "statistical learning," "data mining," and so forth. Each field had slightly different preoccupations: for example, "big data" aimed to take advantage of the fact that computers could store and process data sets of unprecedented size, while "predictive analytics" might be content to work with small data sets but would use them to build models that required a great number of computations. Nevertheless, the fields overlapped considerably and shared the same essence: using the new computational power to extend the reach of statistical prediction.

There are two common and widely advertised ways in which big data (I will use this term in the broad sense, including everything making up the new statistics) is applied to blended learning programs. The first is related to the knowledge tracing form of personalization we discussed earlier. In a sense, knowledge tracing is a prediction problem: a student can be said to know a certain topic whenever she is able to correctly solve problems over that topic. Thus, if we *predict* that a child would solve a problem about pentagons, then we can consider her to know the topic "pentagons." Researchers working on big data in education have developed a wide range of models—each with its own strengths—for accomplishing this task.[16] The idea is that once we have determined via knowledge tracing what topics a student knows, we can more efficiently tailor instruction: we can avoid wasting time on material the student has already mastered or frustrating him with material for which he is missing prerequisites.

The second way is an adaptation of the "à la carte" model from the preceding section, in which *models* now hold the menu: instead of allowing the *teacher* to choose units from a wide range of programs, this approach gives the charge to *statistics*. The idea is to use big data to *predict* how helpful a given unit of material would be for a specific student. If we can successfully make this prediction, then we can pick just the right unit for the right student at the right time, choosing among the hundreds, thousands, or even millions of units developed by various companies or individuals. It is a compelling vision, which is surely part of the reason for its popularity: we imagine a vast virtual bag of online lessons, with inscrutable algorithms combing through them to pluck out *just* the one Johnny needs to at last resolve his long-standing struggle with fractions.

As far as the first use of big data, knowledge tracing, goes, it is most certainly an excellent thing. If we wish to predict which problems a student will solve, there is no better way to do it than with the best available statistical tools. But as far as the second way (which we could term "à la carte, with big data placing the order") goes, it is in my view far less promising—at least for mathematics.

This is for the same reason as for any variety of "à la carte": excellent mathematics education requires great cohesion between topics, and no way of mixing and matching—no matter how technically sophisticated—can compensate for its loss. Frankenstein was a genius, but the beast he assembled from spare parts was nevertheless a monster. A curriculum assembled from spare parts can be nothing different.

There is another reason why big data cannot suffice for curriculum construction. The models only have access to numerical data: how students solve subsequent problems, how long this takes, and so on. This is not too different from the sort of information furnished by testing. But as we saw in the first chapter, this information is not enough to tell apart an excellent curriculum from a middling one. By taking the "à la carte" approach and then transferring control from teachers to machines, we lose the ability to make decisions based on the crucial but unmeasurable qualities of good instruction.

From this it is clear that big data cannot by itself provide excellent content, and thus does not remove the need for instruction modeling. Next we turn to the reverse question: does the instruction modeler have any use for big data? We have already seen one reason why the answer is *yes*: knowledge tracing is a worthwhile complement to instruction modeling, and big data is the best way to do it. But this is not the only reason: there are many more. I will touch on three to illustrate the great potential.

First, researchers have developed so-called "detectors" for student behavior and affect.[17] The best-known researcher in this field is Ryan Baker, a professor of education at the University of Pennsylvania. Baker and his colleagues travel to observe students in blended learning classrooms, recording the students' behavior (on task, off task, conversation with peer, discussion with teacher, etc.) and affect (boredom, concentration, confusion, frustration, etc.). Then they combine this information with the data generated by these same students as they go through the online lessons. From this, they build statistical models that *predict* student behavior and affect. Their models answer the question: "If I went right now to this class and looked at this student, would she be on task? Bored? Confused?" The models have become famous in part because they do this without any video footage of the students: looking only at keystroke data, they can infer with good accuracy the behavior and emotional state of a child! It is a wonderful illustration of the sometimes surprising power of modern statistics.

These "detectors" provide useful information. For example, Reasoning Mind has collaborated with the Baker laboratory to apply them to *implementation*: knowing the classrooms where students are most likely to be bored or off task, Reasoning Mind's implementation coordinators can make better decisions about which teachers need additional help and training.

A second example of the use of big data is *basic research*. Many education researchers can use the insights gleaned from big data to discover new knowledge about how people learn. For example, researchers have used "detectors" to study the relationship between student affect during class and subsequent academic success; it turns out that *boredom* is especially predictive of poor long-term outcomes (such as high school graduation and college matriculation), while *concentration* is predictive of good long-term outcomes; interestingly, frustration and confusion are not as clear-cut, being good in some circumstances and bad in others.[18] This knowledge is clearly useful to designers of programs. For example, it suggests that students who are often bored may need special attention, even beyond what would be given to their confused and frustrated peers.

A third example we will save for the final chapter. It concerns the use of big data to build automated implementation coaches for teachers. In my view, it is one of the major future developments in instructional technology, carrying the potential to radically improve the quality of implementations of comprehensive blended learning programs (such as those developed through instruction modeling).

The Upshot

What have we found from this chapter's survey of some of the most popular approaches to designing blended learning programs? In my view, one major conclusion is that instruction modeling is unique in its ability to guide the creation of excellent *content*. I know of no other approach that offers a comparably consistent, reliable way to do this. Because, as we argued in chapter 6, excellent content is essential for truly excellent blended learning, this is a strong argument for the wider use of instruction modeling.

Our second major conclusion must be that, while none are self-sufficient,* many of these approaches are exceptionally valuable. Even if a designer already employs instruction modeling, she can still do better by also incorporating the other approaches. For example, applied learning science can help design successful interfaces, knowledge tracing can help identify and remedy individual gaps in students' knowledge, and big data can help spot classrooms in need of additional implementation support.

Good blended learning calls for excellent content, interface, and implementation. Instruction modeling sees to the first. Other approaches are invaluable for the remaining two.

* Neither, for that matter, is instruction modeling—as explained in chapter 6.

Chapter 8

How to Conduct Instruction Modeling

This chapter offers a practical guide to instruction modeling. This is of particular relevance to education researchers and developers of blended learning programs. Others may wish to skim or skip this chapter, as it is not required for the remainder of the book.

Being a practical guide, this chapter differs considerably from the others. The others discuss instruction modeling as an impartial idea; while their perspective—as is invariably the case in any work—is colored by the author's experiences, they focus on making specific claims and putting forth factual arguments for these claims' veracity. This chapter, by contrast, treats instruction modeling as a practical craft, offering a practitioner's guidance on how it can be done. I do not claim that the approach described here is the *only* way to do instruction modeling—technically, any approach that seeks to use blended learning to replicate a specific example of offline instruction can be considered to be instruction modeling. What is described in this chapter is merely one way that has been tried and improved over the years. Accordingly, I ask readers to approach it in this spirit: not as a general definition, but as a suggested technique.

Instruction Modeling. George A. Khachatryan, Oxford University Press (2020)

DOI: 10.1093/oso/9780190910709.001.0001

Step 0: Selecting the Tradition to Be Modeled

Before proceeding to use instruction modeling, one should ask the question: "Is it appropriate in the present situation?" As we saw in the previous chapter, instruction modeling has one major strength: done well, it can allow for the systematic creation of high-quality content. It also has one major drawback: it is costly. Thus, instruction modeling is appropriate when a program is to be used by a large number of students (e.g., hundreds of thousands), the content is too large to be perfected through many rounds of iteration (e.g., an entire year-long course), and the quality of instruction is paramount (e.g., a school program as opposed to a corporate training compliance course).

If the situation is deemed to merit instruction modeling, the first thing that must be chosen is the specific instructional tradition to be modeled. Instruction modeling is an ethnographic approach, so "tradition" here means a system of practices employed by a group of people. How large can this group be? In theory, it could be of any size: as small as a single individual or as large as all of the teachers in the world. For example, cognitive scientists Neil Heffernan and Ken Koedinger[1] created Ms. Lindquist, a computer instruction program designed in part to simulate the tutoring practices of an experienced algebra teacher of the same name. At the opposite extreme, Art Graesser's team at the University of Memphis has developed a program called AutoTutor,[2] which simulates the general strategies used by one-on-one tutors in any situation and for any subject. In the broadest possible sense, each of these programs could be considered an example of instruction modeling.

However, if our purpose in conducting instruction modeling is to codify topic-specific instructional practices, then there are drawbacks to using groups that are either very small or very large. With small groups or individuals, the modeler might lose the ability to access the group members over time: they could retire, move away, or lose interest in the modeling project. More grimly, they could die or leave the profession. Additionally, it is difficult in such cases to separate success due to the use of generally strong instructional practices from success attributable to the intangible force of an individual. If a certain teacher is stunningly successful, is this because her instructional practices form an effective, replicable system? Or is it because of her personal charisma, willingness to exert an exceptional degree of effort, or some other quality beyond the reach of the would-be modeler? With a single teacher or a small group, it is hard to know. And in general, systems of effective practices are almost always the product of the sustained efforts of a community; as moral psychologist Jonathan Haidt writes in *The Righteous Mind*,[3] "human beings are 90 percent chimp and 10 percent bee." We may largely be apes, but we also possess a near-magical ability to come together as communities, and these communities can create extraordinary things.

On the other hand, with a very large group of teachers, you are not modeling a content-specific system of practices, but rather universal human reasoning and learning processes—a very valuable thing, but not the aim of instruction modeling as described here. For instance, there does not exist any general "way that the world's teachers teach fractions"—this has varied so radically from country to country and epoch to epoch as to be completely meaningless.

I would suggest that the ideal group size for instruction modeling is large enough to form a community that spans across multiple generations (and in this sense can properly be termed a "tradition"), but small enough that the methods of the group can reasonably be regarded as an interconnected system of content-specific practices. For example, "Russian mathematics teachers using the collection of curricula developed during the Kolmogorov reforms of the 1970s" forms such a group.

Besides finding the right group size, what else matters in identifying a tradition? First, the tradition should be of high instructional quality—after all, in practical applications of instruction modeling, this is what makes a tradition worth modeling in the first place! Determining the quality of a tradition is no simple task, but the first two chapters of this book provide a guide. Second, the tradition should be as systematized as possible. The more consistency and structure a tradition has, the easier it is to model—and the greater confidence the model can have that it indeed contains a wealth of concrete practices. For example, Japanese mathematics instruction is highly systematized: courses and very structured, and teachers have created and published countless example lesson plans for every topic in the curriculum.[4] By contrast, American history instruction varies greatly from teacher to teacher and school to school, making it a more difficult tradition to truly "model." Third (and last), the tradition should lend itself to automation as much as possible. For example, some instructional traditions are heavily student-guided, with the teacher adapting flexibly to decisions made by students in the course of each lesson; when used appropriately, this approach can be highly effective—but it is considerably harder to recreate on a computer than a more prescriptive method. A modeler who chooses such a tradition might develop an excellent understanding of the practices, but then find herself at a loss when it comes to implementing them in an actual blended learning program.

Step 1: Creating a Generic Model (e.g., via the Goals-Practices Method)

Any instructional tradition will have practices at two levels: *content-independent* and *content-specific*. For example, elementary school Russian mathematics teachers will generally begin every lesson with a set of warm-up exercises, often focused on mental math. This is a content-independent practice, since it applies across topics and grade levels. When teaching whole number addition and subtraction in the early grades, these same teachers will emphasize the "decomposition" of a number: i.e., writing 7 as $1+6$, $2+5$, and $3+4$; thus, students will learn to add 8 and 7 via decomposition, writing $8+7=8+2+5=10+5=15$. This is a content-specific practice.

The first step of instruction modeling (once the tradition to be modeled has been selected) is to create a *generic model*—i.e., a detailed description of the *content-independent* practices in the tradition. These practices form the general framework within which teachers in the tradition operate. The generic model answers the questions that transcend any particular lesson: What are the typical ways in which lessons are structured? What types of activities do teachers employ, and how is each carried out? What do students typically do during each lesson? What is the approach to homework? How do students normally interact with the teacher and with one another?

A generic model should be created before individual models for every lesson are made. The generic model is the most difficult step in the instruction modeling process, as the modeler begins with a blank slate. There is nothing to guide the modeler in determining which specific observed practices are important to the tradition, and which are accidental or idiosyncratic to a particular teacher. Modeling a specific lesson is easier, since the generic model is already in place and provides context; but no such context exists when the generic model itself is to be created.

What tools does the instruction modeler have for creating the generic model? There are only two: (1) interviews with teachers* and (2) observations of lessons. Each of these sources provides something the other does not: interviews alone reveal teachers' *ways of thinking* about designing lesson plans, delivering lessons, and responding to students' answers and behaviors. However, observations are also valuable, since they give the modeler a real, direct sense of how lessons in the tradition look and feel. There is a more direct way in which the two sources reinforce each other: observations often provide fodder for interview questions (e.g., "I noticed you were cold-calling students. Why?") while interviews often provide contextual knowledge that renders classroom observations more meaningful.

In my experience, despite the value of observations, interviews remain by far the modeler's richest source of insight. After all, they provide direct access to teachers' decision-making process, along with the great flexibility of being able to dig into whatever question you would like to explore. In a classroom observation, by contrast, you can only see what happens on that particular day.

However, interviews have one significant limitation: they are only as good as the questions asked. Artificial intelligence researchers have long understood that expert knowledge (and teaching expertise is no exception) is not neatly organized.[5] You cannot simply ask teachers to present their overall teaching method and expect a useful reply. Expert teachers take some aspects of their practice for granted, unaware that they are significant and specific to their tradition; for instance, Russian mathematics teachers may assume that all teachers begin each lesson with a mental mathematics warm-up, and therefore this practice requires no comment. In general, expert knowledge is not like a neat textbook with a clear table of contents; rather, it is a messy jumble of ways of thinking, with many complex interdependencies between different ideas but no unifying structure.

* And potentially students—although this is only a supporting source of information, since students do not reveal teachers' instructional decision-making processes, only directly observable classroom activities.

This makes it especially likely that a modeler will commit sins of omission: building a generic model that accurately describes certain practices, but underemphasizes or entirely omits other equally important ones. In chapter 3, we saw several examples of such omissions in the design of Genie 2, subsequently remedied in Edifice.

How can sins of omission be avoided? The best approach I know is the *goals-activities method*. In this technique, the modeler seeks to construct a list of instructional activities a teacher uses, a list of pedagogical goals he pursues, and a mapping between the two. For example, upon learning that most lessons in Russian mathematics classrooms start with a warm-up, the interviewer will ask the teacher to provide a list of goals this warm-up serves to achieve; these goals might include practicing mental mathematics, developing automatic skills for calculating, refocusing students on mathematics (to distract them from whatever they were thinking about prior to the start of the class), and many others. Then the interviewer can ask for each of these goals: "What *other* activities do you use to achieve this goal?"

The goals-activities method exploits the messy, cross-linked nature of expert knowledge. After all, if an instructional activity is important, it usually serves many goals; if an instructional goal is important, it is usually supported by many activities. The goals-activities method uses this many-to-many relationship, traveling along the links in the expert teacher's knowledge structure to explore and uncover as much of it as possible (see Figure 8.1).

In the end, what is a generic model? It is a written document, perhaps between fifty and one hundred pages in length, describing in detail the reasoning and behavior of teachers within the tradition being modeled. There is no standard structure for the document, but I personally prefer a two-part structure: a brief (two- or three -page) summary of the tradition, then Part I with a section for each major instructional goal (describing the goal and listing the activities that support it), and finally Part II with a detailed section for each instructional activity (giving as complete as possible a description of how it is carried out).

As we will see next, this document forms the foundation for everything that follows. While creating it takes time, a detailed and complete generic model creates a scaffold for all subsequent instruction modeling work.

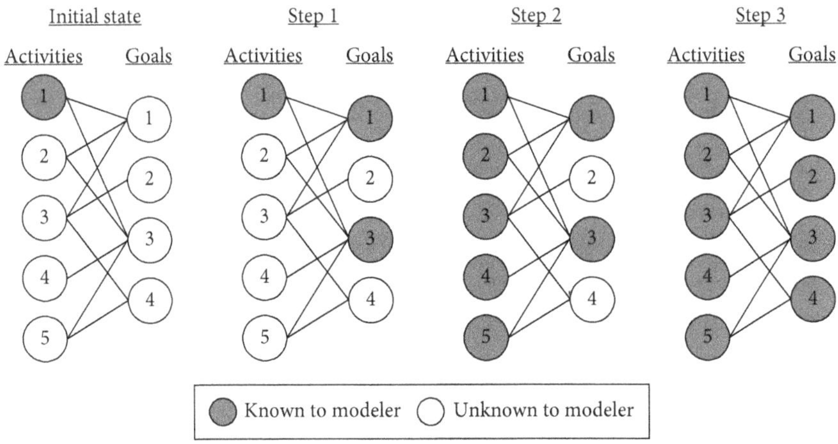

FIGURE 8.1 An illustrative example of the activities-goals method. The modeler initially knows only about Activity #1, but in three steps ends up uncovering the entire knowledge structure.

Step 2: Designing a Platform

The next step is to design the online platform through which the digital instruction will be delivered. This needs to be done *before* individual lessons can be modeled, since the platform will impose constraints on what is possible.

This design has three main goals. First, it should allow for faithful modeling of the tradition—i.e., it should make it possible to create and deliver online lessons that successfully recreate the essential instructional experiences of the tradition being modeled. Second, it should conform with the recommendations of cognitive psychology; as we saw in chapters 4 and 7, instruction modeling by itself does not guarantee the creation of cognitively adequate learning interfaces, and cognitive psychology is essential in filling this gap. And third, the design should be as simple as possible: simple for users, simple to build, and simple to furnish with lessons.

The first goal—enabling faithful modeling—is challenging and requires subtlety of understanding of the generic model developed in the preceding step. For example, assume we are modeling a tradition of literature instruction. We plan to design a platform in which students read samples of text and then answer a series of multiple-choice questions. Will it be adequate? This depends on the tradition we are modeling. Does it require students to hear detailed commentary on texts from the teacher? If so, providing this commentary in a written form may overwhelm students and fail to adequately model the tradition. Does the teacher project text excerpts and mark them up to indicate key passages? If we hope to model this, our platform must support a similar process. Is it important for a student to hear discussions among peers, and perhaps even to actively participate in them? If so, we would be lying to ourselves if we thought a simple sequence of text screens could have the same effect.

In seeking to design a platform to faithfully model the tradition, there are two main pitfalls to avoid. The first is to attempt skirting the issue altogether by designing a "universal" platform, a sort of online swiss army knife capable of modeling anything. As Ezra Cornell said upon founding his eponymous university, "I would found an institution where any person can find instruction in any study." While it worked for him in the matter of universities, this precept is ruinous when applied to the development of online platforms. A good online platform gives users a consistent, seamless experience, allowing them to focus on the content of the instruction; by contrast, if every screen is designed from scratch with no constraints, the user will experience the opposite—a cacophony of different question formats, presentation modes, and widgets to defy the imagination. Furthermore, such "universal" platforms serve to drive up production costs: with each item of content programmed, filmed, or otherwise authored from scratch, there is nothing to streamline the creation process. I have seen firsthand such platforms; they are often the product of well-meaning designers, wishing altruistically to give content authors freedom, seeking to avoid internal disputes about platform functions, but lacking a strong vision for the overall instructional experience to be offered to students using the platform.

The second pitfall in platform design is to carelessly assume that online experiences will adequately reproduce offline ones, even when they are missing some significant features of the offline originals. For example, we might assume that text is a suitable replacement for voice, that multiple choice questions are a suitable replacement for open-ended ones, or that narration is a suitable replacement for discussion. Of course, we cannot fully replicate every aspect of offline instruction online, so some sacrifices must be made; but we must make these sacrifices consciously, using our judgment (and knowledge of cognitive psychology) to determine when they are harmless and when they are severe.

Here, there is an invariable trade-off with the third goal of platform design: simplicity. Richer modeling often comes at the cost of greater complexity: more sophisticated algorithms, greater use of multimedia, and increased interactivity of user questions. Resolving these trade-offs is the art of design, requiring an understanding of the generic model, learning science, software usability, and the practicalities of development costs.

Simplicity in general is the easiest goal to miss. It is tempting to add features. The inexperienced designer thinks: "What's the harm? If users don't like it, they just won't use it." But each feature invariably increases the complexity of the interface, even if only a little; each feature distracts developer resources from polishing and perfecting the truly core components of the platform; and each feature imposes one more design constraint, as interfaces must now be made to accommodate it.

How can the instruction modeler avoid the problem of complexity? The first step is establish that there can be only three legitimate sources of features: first, a feature is legitimate if it enables modeling an experience that is critical in the generic model and that otherwise cannot be modeled; second, a feature is legitimate if it is necessary for conformance with the consensus recommendations* of learning science; and third, a feature is legitimate if it is a direct response to user needs as demonstrated through usability testing or actual platform use. Any feature that does not come from one of these sources should be disqualified on sight. Even among those which do hail from one of these sources, great care should be taken to prioritize. The strongest case can be made for features that come from multiple sources.

For example, Russian mathematics teachers prefer to write very terse worked examples, avoiding long sentences and instead showing problem solutions as numbered steps with formulas in each one. They provide more detailed commentary verbally. For an instruction

* The phrase "consensus recommendations" is significant. Almost any feature can be justified by *some* study in *some* context. The designer should take care to give weight proportionately to the evidence.

modeler, this suggests the need for a multimedia interface: showing worked examples through text, giving a voiceover with further comments, and perhaps providing some animation to point out each step as the voice discusses it. Such an interface is certainly more complex than a simple text screen. But it is also supported by learning science: as we saw in chapter 4, Mayer's cognitive theory of multimedia learning predicts that it would lead to better learning. Furthermore, online lessons that are heavy on text can overwhelm students and yield student and teacher complaints—as indeed they did in some lessons in the sixth grade course of Reasoning Mind's Foundations program. Thus, in this case, all *three* legitimate sources of features speak strongly in favor of the more complex multimedia approach. But this is situation-specific: if we were modeling a different instructional tradition, in another subject, at another age level, then it is entirely conceivable that not all—or even none—of these sources would give strong reasons to use this sort of multimedia interface.

As an example of a more complicated case, take "adaptive learning." Any developer of computer instructional programs feels the temptation to build in adaptive behavior, allowing the program to adjust its instruction to the individual characteristics of the student. The program might adjust the examples to fit students' individual interests, or diagnose gaps in prior knowledge, or simply give more practice when students are struggling with a topic. In this case, the decision is far less clear. To begin with, most adaptive behaviors are extremely complex to implement, calling for complicated software algorithms and often requiring multiple versions of content (since different students are to receive different treatments). Furthermore, the scientific case for adaptive learning is mixed: some forms (e.g., the notion that different students have different "learning styles") have been outright debunked,[6] while others (e.g., the idea of identifying gaps in prior knowledge and filling them in) are well- supported by the evidence.[7] Human tutoring is famously effective, and is often held up as an example of the effectiveness of adaptive learning; however, scholars have called into question whether adaptive behavior is indeed the cause of tutoring's success.[8] Finally, from a modeling perspective, adaptive learning can be irrelevant or even contradictory to a model. For example, Russian mathematics teachers do not greatly vary their style to fit individual student needs not because they are opposed to this, but simply because they teach whole classes. In this instance, implementing adaptive learning would *reduce* the faithfulness of the modeling; if key aspects of the teachers' practices end up being excluded in the process, this would yield a less faithful—and possibly less instructionally effective—model. The modeler has a choice: model faithfully but forget the adaptive behavior, or implement adaptive behavior but risk compromising the model. There is also a compromise (of which I am personally quite fond), in which students take a nonadaptive course, but are rerouted to an adaptive tutorial module in cases when they struggle.

Between these trade-offs and the inherent difficulty of building a computer lesson to simulate an offline one, there will invariably be situations when modeling is impossible or impractical. In the example of Russian mathematics instruction, teachers call students to the front of the class to present their solutions to problems on the board (an activity called "work at the board"); the teacher (and even peers) then questions the student about the solution, probing it for correctness and completeness. It is possible to simulate the experience of *watching* a student being called to the board, but far harder—if not altogether impossible—to build a genuinely faithful simulation of the experience of *being* at the board.

What can the modeler do in such cases? There are two possibilities. First, you can accept defeat, adding the experience to the mental tally of modeling failures. If the tally is not too long, you can still hope for an outcome that is on the whole successful. But if the battle is too important to concede, there is a second option: you can plan to train classroom teachers to deliver the experience offline. Indeed, this is the *blended* aspect of blended learning: the computer has a role, but so does the teacher. For instance, in the example of "work at the board," we can model the experience by training classroom teachers to pull students aside for one-on-one or small-group discussions of problem solutions, with students presenting their work for critique by the teacher.

This might seem like an easy escape hatch: if you can have classroom teachers deliver any experience, why bother with modeling much online in the first place? The answer has to do with the difficulty of implementation. You can design offline activities in any quantity, but teachers will only succeed in proportion to the training and administrative support that is provided. And the larger the scale of the program's use, the harder it will be to provide these things.

The more activities are asked of teachers, the lower the consistency and fidelity of each one will be. For this reason, instruction modelers should be cautious both to keep the list of offline activities small, and to devote considerable thought to creating adequate training and support materials for teachers undertaking these activities.

Step 3: Modeling Content

The final step of the process is to model the actual course content: the substance of each lesson. In a sense, all of the previous steps of instruction modeling serve only as the setup for this step. After all, as chapters 6 and 7 argued, instruction modeling is useful precisely for its value in creating excellent content.

There are two approaches that the instruction modeler can take at this stage. In the robust approach, each lesson can be treated as a miniature version of the generic model: the instruction modeler conducts interviews with expert teachers to understand how they conduct a given lesson offline, and then creates online lessons on the platform designed to recreate the same instructional experiences. For example, in building Reasoning Mind's Edifice program, expert teachers wrote scripts (typically several dozen pages per lesson) describing how they taught it offline; then, for each lesson, an instruction modeler conducted phone interviews with the teachers to clarify and understand it. Finally, the instruction modeler would create the online lesson designed to mimic the offline one.

In a simplified approach, the two steps are collapsed, with expert teachers asked to create content directly. This approach was taken, for example, in Reasoning Mind's Foundations program. Teachers were given guidelines (e.g., on the technological constraints of the online platform) and asked to write documents with scripts for the online lessons. These scripts were subsequently heavily edited by curriculum developers (to ensure they were detailed and implementable) and then sent along to the content creation team for programming.

The simplified approach is generally less expensive. For example, it requires half the documentation: the robust approach requires a document describing the offline lesson *and* one describing the online one, whereas the simplified approach needs only the second. However, the robust approach yields considerably better modeling. This is because expert teachers—like any other kind of expert—are experts in their discipline (viz., teaching), *not* in modeling. They are trained to reason expertly in authentic situations—i.e., standing in front of a classroom of students—but not to design computer programs to do the same. Anyone who has done instruction modeling (or "knowledge engineering," as it is called in a general computer science context) has seen firsthand how shocking this gap can be. For example, a teacher might ask students to solve a problem in small groups; when charged with creating an online version of this lesson, the same teacher might simply propose a text screen with the problem's solution. When asked why the difference, she might say: "How would you put a small group discussion in a computer lesson? I can't see a way, so I just put the screen." Worse still, teachers might deliberately throw out their expertise, bedazzled by the allure of computers. A teacher might propose a complicated interactive widget with no connection to their classroom practice; when asked about it, he might say: "I can't do this in my regular classroom, but with a computer it's possible! So I took the opportunity." Of course, unlike whatever the teacher did in the offline classroom, this new widget is entirely untested, and might have little actual instructional value. In sum, experts often undervalue their own expertise, are unaware that some things they do "out of habit" are actually significant practices perfected over time by their tradition, or put insufficient thought into how their offline practices might be cleverly modeled online. This is why the robust approach to content modeling—with separate individuals explicitly trained as instruction modelers—works much better when it can be afforded. Instead of asking expert teachers, "How would you have a computer teach this lesson?" they are asked a question that is truer to their authentic practice and expertise: "How do you teach this lesson, and why?"

Regardless of which approach (robust or simplified) is taken, it is very important to create good documentation templates. Instruction modeling is heavy on lesson-by-lesson documentation,* so the format of documents plays a major role in the quality and efficiency of content creation. Documents describing offline lessons as well as ones describing online lessons (to be provided to programmers and others

* At least in the traditional approach. Chapter 9 will argue that rapid authorship has considerable potential for reducing the costs of instruction modeling. However, the challenge there will be find ways to preserve the fidelity of the modeling.

charged with producing actual content) should be quick to write, easy to read, and should make it easy for the reader to visualize the online lesson being proposed. One important decision is whether to include *mockups*—drawings of what screens will look like—in documents. On one hand, such drawings take time to create. On the other hand, they make it much easier for readers (and, in fact, writers!) to mentally picture the proposed lesson. By contrast, text-only documents require the reader to mentally build up the lesson, keeping in the mind an image of the screen. Such documentation templates invariably lead to online lessons in which essential information is removed from the screen too early (causing subsequent learner confusion), or unnecessary content remains (cluttering the screen) well past its time. In a way, documentation built around a sequence of mockup screens is almost a "prototype" of the online lesson. For this reason, my experience has suggested that investing time in mockups pays for itself by reducing the amount of rework once content is completed.

Can Instruction Modeling Be Agile?

Any reader is likely to notice that this process is long and cumbersome: instruction modeling, as described here, is a lot of work. Can this be avoided? To an extent, yes—but at a cost. One potential simplification is to skip the formal creation of a generic model. Another is to skip the creation of written lesson models, instead asking expert teachers to directly create documents describing online lessons. Each of these simplifications can save time, but each comes at a cost in fidelity of modeling. Instruction modelers should make the decision in each case based on the available resources and the relative importance of content quality.

Readers familiar with software design may have another question: why does this process appear so much like a "waterfall" instead of an "agile" approach? The waterfall approach is the classical method of software development. First, users are interviewed and an overall vision for a software product is created. Then, documents are written describing how the software should behave (called "functional specifications" or "functional requirements"). After this, programmers write the software code. Next, testers work with the software to remove bugs. Finally, the software is rolled out to actual users. The approach is called the waterfall method because it resembles a sequence of discrete, irreversible steps, like a cascade waterfall (Figure 8.2).

In the last two decades, the waterfall method has fallen out of fashion, replaced by an approach called "agile development" (or just "agile"). Adherents of agile criticize the waterfall method as inefficient and ill-suited to the creation of easy-to-use software. While the waterfall method appears neat and orderly, the rigid sequence of steps causes problems: no one can actually see and work with the product until a great deal of effort has been expended to write documentation and code large pieces of the software, and users do not

FIGURE 8.2 A cascade in Ithaca, New York. In the "waterfall method," software is created through a rigid sequence of discrete steps, from design to testing.

actually experience it (and provide their feedback) until the full project is completed. In a waterfall approach, developers risk spending years only to discover at the end that their approach does not truly meet users' needs, or creates insurmountable usability problems. In an agile approach, by contrast, designers intermingle the steps. They seek to create a functioning bare-bones version of the product (called a "minimal viable product") as quickly as possible, testing it themselves and with users. Then they iteratively improve the product. Thus, the development of specifications, coding, and testing all take place concurrently and reinforce one another. The agile approach has been so successful that today's leading software companies near-universally use it to the exclusion of waterfall methods.

Instruction modeling, as described in this chapter, is clearly a process along the lines of the waterfall method. The software platform is designed before individual lessons. Each lesson is modeled first, then documented, and finally created. Thus, the approach is susceptible to the same criticisms (and problems!) as any kind of software development following the waterfall method.

To some extent, this is unavoidable: instruction modeling is about replicating an existing way of teaching, and thus the "design" is already in place. Furthermore, you cannot develop dozens of individual lessons before the online learning platform has been put in place. This certainly is a drawback to instruction modeling, as programs developed using this approach can meet with variable success with users.

The risk can be somewhat reduced by introducing as many elements of agile development as possible. First, the software platform can be created in an agile manner, iteratively creating increasingly sophisticated versions and testing with users each time. Second, one or two complete lessons can be created concurrently with the platform, offering a "preview" of subsequent steps and allowing them to be tested early in the process. Third, in cases when some lesson components are expected to occur often, prototypes can be made and tested early in the process (e.g., prototyping and testing any unusual interfaces for providing answers, such as tools for entering formulas or making geometric constructions). Fourth, short versions of courses can be created and tested with users before resource commitments are made to the creation of large ones using the same platform. And fifth, each lesson can be developed in a more agile way by writing documentation that makes it easy to visualize the lesson before it is actually created—e.g., by showing a sequence of drawings of screens to be shown over the course of the lesson.

In conclusion, the instruction modeling process described here can be made considerably more agile than it might appear. Nevertheless, accomplishing this can be challenging and requires deliberate effort on the part of instruction modelers. Just as with any type of software development, agile steps—such creating prototypes and conducing usability tests with users—appear at first glance to be costly and to delay development, but in the long term serve to expedite the process and reduce the risk of product rejection.

Chapter 9

The Future of Instruction Modeling

IN CHAPTERS 3, 4, AND 5, WE SAW THE RAPID PROGRESSION of Reasoning Mind's blended learning programs, with each program yielding surprising lessons that informed the next. Most of the developments were directly related to instruction modeling: each time, new ways were found to more faithfully model the experiences of students in exceptional mathematics classrooms.

This journey is far from complete. An engineer who designs computers will never say, "This model has sufficient speed for any purpose. I think our work is finished." The same is true of instruction modeling, and more broadly of blended learning design of any type. There is no end to potential improvement, and I cannot imagine a world in which we will collectively judge our children to be sufficiently well-educated, and in no need of better instruction.

This chapter discusses what I regard as the most exciting future directions for instruction modeling. Each is ambitious and grand in scale. Each would in itself constitute a great leap forward. And each is entirely achievable with methods and technologies that are already in existence. Collectively, these possibilities illustrate just how far instruction modeling is from its full potential. To me, this is reason for

Instruction Modeling. George A. Khachatryan, Oxford University Press (2020)

DOI: 10.1093/oso/9780190910709.001.0001

great optimism: something that is already successful can be greatly improved. It will take more than one decade. But I believe that the cause is worthy of the effort.

Richer Interfaces: Natural Language and Collaboration

We have already discussed the importance of the *interface* in a blended learning program. By "interface," we mean more than just the software screen: we include *all* of the ways in which the substance of instruction is delivered. For example, in a conventional classroom, a paragraph of text read aloud by the teacher would carry the same substance (i.e., the same *content*) as the same paragraph shown as text on the blackboard, but would be delivered through a very different *interface*.

One of the most important consequences of Mayer's cognitive theory of multimedia learning (CTML), which we discussed in chapter 4, is that the interface plays an important role in whether students make an effort to actively think about the content presented. Identical content offered through subtly different interfaces can yield radically different experiences for students.

Many psychologists besides Mayer have investigated this question. Perhaps the most influential has been Michelene Chi of Arizona State University. Chi's most famous work has concerned "self-explanation," a strategy for understanding material.[1] In "self-explanation," the student is prompted to explain a text (or video, sound recording, etc.) out loud to herself. For example, we might ask a student to read a textbook chapter; after every paragraph, we would ask her to explain that paragraph aloud. We would provide no response to these explanations, simply asking the student to move on after each one. Chi's finding was that this surprisingly simple practice—i.e., prompting students to verbally explain material to themselves—had an extremely large effect on how well they learned the material. This became known as the "self-explanation effect," and has been the subject of countless follow-up studies.

Of course, self-explanation as a learning strategy is not a recent "invention." For example, when she was a fourth-grade student, my wife took up the practice of explaining her school lessons to her pillows. She subsequently went on to become a scientist, so the method must have worked. We do not know what became of the pillows.

But while self-explanation itself was old, the discovery that merely prompting students to self-explain was sufficient to have a large effect was new. Mere prompting could take self-explanation—previously the province of the exceptionally conscientious—and bring its benefits to the huddled masses.

Chi subsequently proposed the ICAP (interactive-constructive-active-passive) framework, which placed self-explanation into a broader context.[2] This framework classifies learning experiences (or, in our language, *interfaces*—since ICAP is agnostic as to the substance of the instruction) into four categories, given in descending order of desirability (Figure 9.1). The bottom category, "passive," includes those situations in which there is no evidence that students are actually interacting with the content. For example, students who are silently attending a lecture, or quietly looking through a textbook, are engaged in "passive" activities. A step above, "active" activities are those in which students show physical evidence of interaction with the material: taking notes at a lecture, underlining as they read a book, answering numerical or multiple-choice questions in an online lesson. The next step consists of "constructive" activities: these are activities like self-explanation, in which students

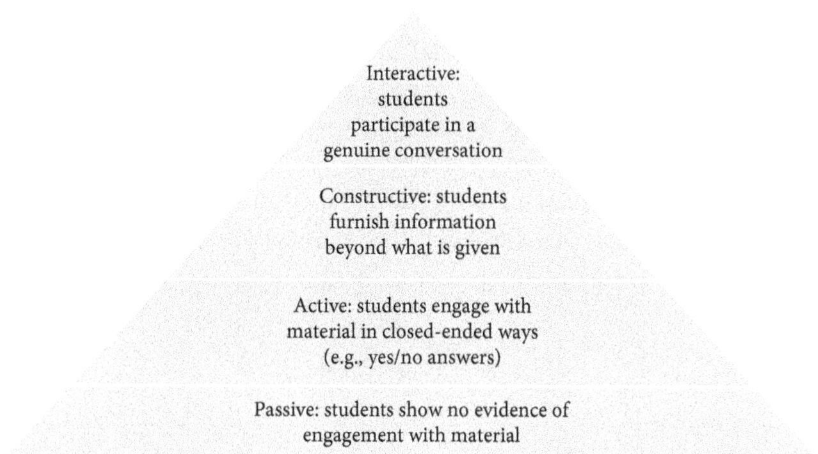

FIGURE 9.1 Chi's ICAP (interactive-constructive-active-passive) framework for student learning experiences.

provide *open-ended* answers (e.g., speaking or writing) that furnish information beyond what has been explicitly given. For example, a student who is repeating verbatim or merely paraphrasing from a textbook will not be judged to be engaged in a "constructive" activity, in contrast to a student who provides good self-explanations (i.e., ones that actually comment on the material instead of merely repeating it). Finally, at the peak we find "interactive" activities: those in which students not only contribute constructively, but actually engage in a conversation—i.e., receive constructive feedback to their constructive contributions. A classic example of an "interactive" activity would be a one-on-one tutoring session, and another example would be a discussion between three students working together to make sense of a problem or book passage.*

There is considerable scientific evidence that each category is superior to the ones below it: if everything else is held constant (as it can be in the laboratory), then "active" activities tend to yield better learning than "passive" ones, "constructive" better than "active" ones, and "interactive" better than "constructive" ones. The only questionable link is the last one: of the studies that have directly tested it, some surprisingly found the "constructive" activities to outperform the "interactive" ones.[3] Chi explains this (in my view, convincingly) as a consequence of the difficulty of actually obtaining true "interactive" activities: in those experiments where "interactive" activities did poorly, the experimental conditions had failed to obtain genuine interaction. For example, one experiment asked students to explain textbook material to one another, but devolved into students simply reading the passages aloud or paraphrasing; understandably, the students ended up learning *worse* than students in a comparable "constructive" condition who were asked to individually self-explain the same material. One-on-one tutoring does not suffer from this problem, but of course is very expensive. On balance, the evidence seems to suggest that "interactive" conditions are the best, but suffer from a "stability" problem: if they are not designed

* Note that ICAP is formulated in terms of directly observable student behaviors. Mayer's CTML, by contrast, deduces recommendations from an underlying (not directly observable) theory of the mind. These are two distinct schools of thought. Psychologists might debate which is preferable; each has undeniable merits. We designers, meanwhile, are omnivores.

just right, the result might drop not one, but two steps in the ICAP framework.

Where are most blended learning programs in Chi's framework? The most common position is somewhere between "passive" and "active." For example, some programs show videos ("passive") and follow up with problems ("active"). Others focus on problems ("active") and give explanations of misunderstood concepts through lengthy text passages ("passive"). This is not too different from the majority of conventional classrooms, in which students spend most of their time passively listening and some of their time on simple, procedural tasks. Reasoning Mind's programs do somewhat better: there is almost no "passive" content (since all explanations are presented with frequent interspersed questions to ensure students' attention), and there is a considerable further effort (via the implementation support) to encourage teachers to conduct small-group and individual tutorial sessions with students ("interactive"). However, this still leaves the bulk of students' time in the "active" category, and invites the question of whether there might not be some way to upgrade the online experience into the "constructive" or even the "interactive" province.

I believe that there are two exciting ways this can be brought about. The first is through *natural language processing*: the ability of computers to make sense of written language. If a computer could understand text typed or spoken by students, then an online lesson could ask the same questions a tutor or teacher asks: "Why did you do it this way? What do you think our next step will be? What's wrong with this answer? What does 'denominator' mean?" Today, with the appearance of mobile devices that take arbitrary verbal commands (like the famous Siri), this seems less like science fiction than before. Nevertheless, computers that convert written text into genuine meaning are nowhere in sight. A search engine can do an excellent job of recommending websites based on typed text, but it cannot carry on a sensible conversation. Thus, expecting a computer to accurately evaluate a student's answer to an open-ended question is too optimistic.

Education researchers have found some ingenious workarounds. One example is the work of Danielle McNamara, a psychologist at Arizona State University. McNamara and her colleagues have developed iSTART, an online program that displays successive passages from a book and prompts students to self-explain each one by typing into a text box.[4] The program cannot make sense of students' responses, but it *can* use statistical methods (specifically, a method called *latent semantic analysis*, or LSA) to estimate their similarity to the original passage. If a student's response is too similar, then it is most likely only a repetition or paraphrase. If too different, then it is most likely nonsense. But if it is in the middle, then it is quite likely to be a reasonable self-explanation. By rewarding students for entering responses that pass this check, iSTART successfully elicits genuine self-explanations.

Take, for example, this stanza from Fitzgerald's adaptation of the *Rubaiyat* of Omar Khayyam:

And that inverted Bowl we call The Sky,
Whereunder crawling coop't we live and die,
Lift not thy hands to it for help—for It
Rolls impotently on as Thou or I.

A response that only repeated words or phrases from the passage would be marked as too similar. A response that was irrelevant (e.g., "Zebra zebra giraffe gefiltefish") would be marked as too different. But a sensible self-explanation—e.g., "The gesture of lifting hands up means asking for help from the divine, represented by the inverted bowl of the sky"—would pass the test. Of course, this approach does not attempt to assess the *correctness* of the response: "The inverted bowl had been full of cereal, and now the speaker cannot lift his hands because he lacks the requisite nutrients" might pass just as easily. But this is mostly beside the point: the goal is only to elicit good-faith attempts, since this by itself is enough to upgrade the reading of the passages from "passive" all the way to "constructive" in the ICAP framework.

A second group of education researchers employing natural language processing is based at the University of Memphis. Arthur Graesser, Xiangen Hu, and their colleagues have developed a program called AutoTutor, which simulates a one-on-one tutorial.[5] AutoTutor asks questions and accepts typed student responses; it grades the responses by statistically comparing them to some stored samples of correct ones. For example, the program might ask: "To what are the hands in the passage being lifted?" The student's typed response is statistically compared to two stored correct ones: "the sky" and "the inverted bowl." If the answer scores high on similarity* to one of these, it is counted as correct. As you might suppose, this method of grading is usually right but not infrequently wrong; accordingly, AutoTutor takes great care to avoid categorical pronouncements. If a student's reply scores low for similarity to the correct answers, AutoTutor might say: "I see . . . in my opinion, the hands are being lifted up to the sky," thus implying—but not directly asserting—that the student's answer was deficient. Like McNamara's iStart, AutoTutor aims above all to obtain a good-faith effort from students, pushing them from "passive" into "constructive"—or perhaps even "interactive"—territory.

What are the implications for instruction modeling? Approaches similar to those of iStart and AutoTutor could be used in programs developed using instruction modeling. For example, an open-ended question asked by a teacher is usually modeled in Edifice as a multiple-choice question, which would fall into the "active" category in the ICAP framework. In future programs, it might be possible to ask the question and evaluate students' typed responses, thus upgrading into the "constructive" realm.

Besides natural language processing, *collaborative learning*[6] might someday be used in instruction modeling programs. For example, two students might take a lesson concurrently, either by sharing a computer or by connecting remotely (as in an online multiplayer game).[7] The students might be prompted to speak with each other, answering questions or discussing various aspects of the lesson. Done well, this might upgrade the student experience all the way to "interactive."

Natural language processing and collaborative learning would improve the instructional quality of the *interface*, but also would allow for more accurate modeling. After all, teachers in conventional classrooms ask open-ended questions, and students in conventional classrooms interact with their peers; these experiences could now be modeled more faithfully. This is a familiar state of affairs (cf. the application of Mayer's CTML in Edifice, as described in chapter 5): in the matter of the *interface*, learning science is a far sturdier guide

* Thanks to latent semantic analysis, this comparison takes into account some of the structure of the English language. For example, "the heavens" might be judged as very statistically similar to "the sky," just as a web search for "films" would give very similar search results to a search for "movies."

than instruction modeling, but improvements done in the name of learning science nevertheless often improve the fidelity of modeling.

Beyond Mathematics

Introducing natural language processing and collaborative learning into instruction modeling programs will require a great deal of effort. As with any major design change, there will be unintended consequences, and years of trial and error will be needed to iron things out. Nevertheless, I believe that it is well worth the effort: these techniques would certainly strengthen mathematics programs, but—perhaps even more consequentially—they would open up enormous new vistas for instruction modeling, allowing the approach to be applied to subjects in the sciences and humanities.

After all, mathematics is unique in the degree to which it is amenable to closed-ended questions. A program like Edifice or Foundations, relying exclusively on closed-ended (numerical, multiple choice, or similar) questions, can still faithfully model the majority of the experiences that students have in a good conventional mathematics classroom. But consider a subject like literature—or even biology—and an entirely different picture emerges. A good literature class cannot be taught primarily through closed-ended questions; we can only answer which river Achilles was dipped in so many times until we acquire an abiding, lifelong hatred for the written word. In a class that teaches literature meaningfully, genuine discussion must predominate.[8]

Natural language processing can bring this possibility closer. Imagine, for example, a program like Edifice equipped with natural language processing tools similar to those in AutoTutor. Recall that in Edifice, a group of three characters—a tutor and two students—carry on scripted conversations with one another and with the actual student taking the lesson. To model a literary discussion, we could have the characters recreate the discussion verbatim, asking the actual student to opine or respond to specific questions through typing. The characters would respond to the student's entry differently depending on how closely it matched certain anticipated (common) answers. Of course, this would not be a complete simulation of actually participating in a literature seminar; however, it would contain many of the right elements: the student would be exposed to literary discussion, would be pushed to offer opinions, and would deepen her understanding of the discussed readings. Of course, if we further added a collaborative learning component—e.g., having two real students attend the same online lesson, prompted to discuss certain issues as appropriate—then we could hope to get even closer to reproducing the original experience.

This approach would not *replace* actual discussion, but it might be *combined* with it to provide a complete education. For example, schools with unlimited resources prefer to teach literature through small-group discussions. This is true of liberal arts colleges, but it is also true of many private high schools; for example, Phillips Exeter Academy educates students at the famous Harkness table, an oval-shaped construction that seats a dozen. But in most of the nation's thirty-some-thousand high schools, literature classes are more likely to hold thirty students. How much genuine discussion is possible in a group of this size? Now imagine applying instruction modeling in this context. Twenty students could take online lessons, which would be designed to emulate the world's best literature seminars; as we have seen, these lessons would already have considerable elements of discussion, showing students daily how books can be read, understood, and talked about. Meanwhile, the remaining ten students could participate in a genuine literary seminar with the teacher. The next day, they would rotate, so over the course of the year each student would spend two-thirds of her time independently working online and one-third engaged in deep small-group discussions. The teacher and students alike would learn from the online lessons how to moderate and constructively participate in literary conversations, and would apply this knowledge in their own seminars. If such a program were well designed and implemented, it is not difficult to imagine the result: students in countless classrooms will finally have access to the sort of literature instruction previously only affordable for the most exclusive private schools and colleges.

Because mathematics is such a well-defined discipline,* it is in some sense the easiest to model online: most claims made in mathematics classes are either right or wrong, and in the majority of cases a

* Though, as we discussed in chapter 2, not so well-defined as many people think: good mathematics instruction relies heavily on discussion, explanation, language, and other such elements going beyond calculation.

computer can accurately distinguish the two. But mathematics is not the only subject that matters in our schools. As instruction modeling broadens its toolkit, it will allow us to faithfully model the world's best instructional traditions in literature—and in biology, chemistry, philosophy, history, and much else.

Expanded Supports for Teachers

As we have seen again and again, good implementation is critical for the instructional success of a blended learning program. Of the three dimensions we identified in chapter 6—content, interface, implementation—the most neglected is content, but implementation is the most difficult. This is because the content and the interface are determined centrally, when a program is developed; meanwhile, the implementation is a battle that must be fought anew each day by every teacher who uses a program. For the battle to be won more often than it is lost, teachers must have solid training and implementation support.

In programs that aim to change the content of instruction—and any program designed using instruction modeling falls into this category by definition—the challenge of implementation is all the greater. Teachers must not only understand how to administer the program (what we termed "facilitation" in chapter 4), but also must learn the new instructional approach well enough to effectively work with students individually and in small groups.

What is the present state of implementation in Reasoning Mind's programs? While in many ways successful,[9] Reasoning Mind's implementation support still has some significant limitations—limitations that are shared by every other existing blended learning program I know of, but that are limitations nonetheless. The first is expense: having support staff personally visit teachers to observe their work is quite possibly the most important ingredient in Reasoning Mind's implementation recipe, but it adds to the program's cost and thus slows expansion. Furthermore, several visits per year are certainly far better than nothing, but still fall short of the ideal: weekly or even daily feedback to help teachers perfect their methods. Finally, there is much more currently in place to help with "facilitation" than with the more difficult task of equipping teachers with a deep understanding of the program's content. This is understandable, since facilitation is what makes the difference between a disastrous and a passable implementation; but of course, deep teacher content knowledge is necessary for further progress.

Is it possible to remove—or at any rate reduce—these three limitations? In my view, new methods arising from big data in education might offer a remarkable solution.

In the previous chapter, we saw how modern statistics (called "big data" and also known by many other loosely synonymous names) could allow blended learning programs to infer some things that ordinarily could only be ascertained by human observation. For example, Ryan Baker and colleagues have developed statistical "detectors" that rely exclusively on student keystroke data to determine whether students are bored, frustrated, confused, off task, and so forth.[10] These detectors are, in essence, automated observers: they do exactly what a human being might do in visiting a classroom, but they do it constantly and at no cost beyond their initial development. It is entirely possible to extend Baker's methods to develop such "detectors" for almost all of the "facilitation" activities needed for a successful implementation—indeed, initial work in this direction has already been carried out through a collaboration between Reasoning Mind and the Baker lab.[11]

Ultimately, it would be possible to build "detectors" that would know whether students were engaged, whether they were appropriately using notebooks, whether the teacher was spending enough time working directly with students, and many other such things. This information could then be used to build an automated "coach" for teachers. At the end of each day, the teacher would see how the implementation fared that day; the coach would celebrate the teacher's successes* and would offer suggestions—in the form of videos or online lessons—for how to remedy any deficiencies. For example, the coach might celebrate the consistency with which students use notebooks, but also note that time on task is lower than the goal; it would suggest to the teacher a twenty-minute online lesson

* Lest you doubt the value of praise served up by a machine, just ask anyone who has worn a Fitbit or similar device: the three-second celebration of the moment when the wearer completes ten thousand steps (which my wife calls a "mini-party") is a highlight of the day for many wearers.

including video interviews with Reasoning Mind teachers who have excelled in keeping students on task and engaged.

This automated "coach" could solve the problems of cost and of observation infrequency. Of course, implementation staff would still be needed, since a personal connection to a supportive human being is important in any complex reform. But much less travel to observe individual teachers would be required, allowing each implementation coordinator to support many more teachers.

The automated "coach" could also help teachers acquire the mathematical knowledge needed to effectively work with individual students and small groups. For example, it might offer short online lessons on mathematical topics through the same interface that offers coaching on facilitation activities. Furthermore, it would free the time of implementation coordinators to focus more on instructional matters in their work with teachers.

Given the importance and difficulty of implementation fidelity, I cannot overstate the contribution that such an automated "coach" might make to improving the quality of blended learning programs. Programs that are comprehensive, and thus challenging to implement—such as ones developed using instruction modeling—stand to benefit the most. But any blended learning program that seeks successful implementation would gain much.

Of course, the prospect of automated daily observation raises a natural concern: what of teachers' privacy? There are two facets to this question: does such monitoring violate teachers' rights? And even if it does not, is it a good idea? First, in the matter of privacy *rights*, I think that few would argue that teachers have a right to professional privacy from their supervisors in the same sense in which students and parents have a right to privacy for their educational data. After all, it is perfectly legal—and widely practiced in every kind of work—for employers to monitor the work of employees, provided this monitoring is confined to the employees' professional activity. Programming managers review their team members' code, call center supervisors listen to operators' phone conversations, and construction superintendents watch workers as they go about their labor.

But the second facet of our question—namely, whether such monitoring is *advisable*—is more complex. Excessive observation, even when perfectly legal, runs the risk of creating a stifling and uncomfortable work environment. For example, many teachers would balk at having their classroom constantly filmed, just as many police officers object (whether rightly or wrongly) to wearing body cameras. But not all observation has this effect: much is neutral or even appreciated. The previously mentioned digital pedometers—such as the Fitbit—offer a ready example of the latter. So our question is really this: which category would the monitoring carried out by the automated "coach" fall into? Would it be a welcome support, or an annoying imposition? This can only be found out by building and testing such a system, but my view is that if done well, it would be seen mostly as a help. After all, much of the objection to video cameras is that they intrude beyond what is professionally relevant: they capture not only professional activities, but also banter with colleagues, the drinking of morning coffee, and so forth. The "detectors" would not have this flaw, as they would report exclusively on professionally relevant measures. They would know whether the teacher kept students on task, but would have no idea of whether she drank one or two cups of coffee, whether she smiled or frowned while reading her e-mails, or whether her watercooler jokes that morning were any good. Thus, the information itself will contain nothing intrinsically intrusive.

The real determinant of how teachers receive such a "coach" will, in my opinion, be how it is used. If administrators use it punitively, then it will be a frustration; if as a source of guidance on improvement, then a welcome service. But this trade-off is nothing new: already, plenty of data is available that holds within itself the same possibilities. Programs record time logged online. They record how far students have progressed. They record which online activities they undertook. Reasoning Mind's experience has shown that when administrators review the data but do not tie it to teacher performance reviews, it is innocuous. But when they go further, setting firm targets and holding teachers to account, the result is higher implementation fidelity but lower teacher morale. But this should come as no surprise: the same trade-off is inherent in using any kind of numerical data for management, whether in education or in any other field.

Rapid Authorship

One limitation of instruction modeling is that it is laborious, and another is that it is inherently ill-suited to departures from existing instructional practice. It takes a great deal of effort to document the practices of expert teachers, a great deal more to plan online lessons that reproduce these practices, and then a third great deal to actually program the lessons. If the course is planned to be different from what the expert teachers have done before, then it runs a much higher risk of turning out badly and needing to be redone—a serious problem when making just one version requires a year of work for several dozen people.

These drawbacks could be mitigated if there were ways to quickly create online lessons in the context of instruction modeling. Even if those lessons were less polished than what current methods produce, they could serve as prototypes—or even as fully functioning courses for schools that care more about the content than about the external appearance of a blended learning program. Thus, an instruction modeling course could be put together rapidly using the quick approach, and then tested and iteratively improved. Once a satisfactory version is reached, the investment could be made to produce a polished version if such a version were deemed necessary.

But is it possible to carry out this sort of "rapid authorship"? In my view, the Edifice program offers a clue. In Edifice lessons, the student sits in on a small group lesson, delivered by a tutor character to two student characters and to the real student. The characters' speech is fully scripted, but they frequently interact with the real student, asking questions and responding to the student's answers. Might not a video recording of a real small-group session, edited to pause every minute and ask the real student a question, achieve a similar effect?

A collection of such videos could be produced with relative ease: simply ask an expert teacher to instruct a group of two students through an online tutoring interface, recording all of the sessions. The teacher would be asked to conduct the lessons exactly in the same way she conducts her regular class. At the end of the year, another person would take the videos and edit them to pause in appropriate spots, ask questions, evaluate the student's answers (perhaps using natural language processing to make freeform text prompts possible), respond appropriately, and then resume the video. The result would be a scrappier, less editable, less carefully designed version of Edifice lessons. But it would also be worlds easier to make.

The tools to create such lessons are not far away. For example, a company called Zaption offered a tool for editing instructional videos to make them interactive. Zaption was bought by a workforce training company and closed its platform, but nevertheless it illustrates the ease of such an approach.

Several questions arise. First, how can this work if the expert teacher does not speak fluent English? In such cases, it could be done in two steps: first the expert teacher's regular classroom is filmed for a full year, and then an English-speaking curriculum developer conducts online group lessons (also recorded) that aim to mimic the expert teacher's lessons. This would introduce some additional error and would be more costly, but would still be many times easier than creating a fully polished course.

Second, why should we have students in the videos at all? Cannot the teacher speak directly to viewers, offering an instructional monologue? Certainly this is possible. Moreover, it would be logistically simpler to make such videos. In some situations, this approach might be just what is needed. However, we covered the importance of dialogue and peer effects in mathematics education (chapter 2), and in other disciplines it is at least as great. Removing dialogue would surely be a detriment to learning. Furthermore, asking a teacher who traditionally works with a class to deliver a monologue will surely distort the result. We saw in chapter 3 that asking teachers to write scripts for online lessons led to significant inaccuracies, and the same would take place with a recorded monologue.

Third, how can we be sure that this approach—even with students in the videos—would yield an accurate model? After all, an online tutorial is quite different from an offline class. The teacher might adjust to the new genre, throwing out her years of experience in the process. This is a very valid concern. My own sense is that some distortion would take place, but it would fall within acceptable bounds. But my own sense is irrelevant: the question is an empirical one, and can be tested by creating such lessons and seeing where they land.

If a method for rapid authorship or prototyping in the context of instruction modeling can be developed, it will carry many outstanding benefits. It will make it possible to apply instruction modeling to subjects beyond mathematics much faster, bringing excellent instruction in those subjects to many more students; it will make it possible to experiment cheaply with different approaches to instruction modeling, building and testing courses based on novel ideas; and it will make it possible to develop courses that depart in certain ways from established practice, testing and iteratively improving them until satisfactory. In other words, it will be a great practical step forward.

Designing Schools for Instruction Modeling

In chapter 5, we discussed the division of the teacher's work in a blended learning classroom into two categories: *facilitation* and *instruction*. Facilitation is the administering of the classroom, while instruction consists in directly educational tasks, such as tutoring an individual student or a small group. This division is not abstract: it is quite obvious whether a teacher is engaged in instruction or facilitation at any given moment.*

In a conventional classroom, facilitation and instruction are combined: the teacher usually instructs the whole class, concurrently watching students to ensure that they are on task and paying attention. In a blended learning classroom, this is no longer possible. The students are at any moment divided into two groups: those at their computers, and those meant to be working directly with the teacher (individually or in a group); the teacher must *facilitate* the work of the first group, and *instruct* the second. A one-headed teacher cannot do both in the same moment.

In theory, this should not present a problem. By establishing clear procedures and expectations, the teacher can ensure that students work productively at their computers without the need for constant supervision. Then, he can spend most of his time on instructional activities, pulling students aside for tutoring and checking up on the remainder of the class every few minutes. In the best classrooms, this is precisely what takes place.

But most classrooms fall at least somewhat short of this ideal. To begin with, the ideal calls for very strong classroom management

* The reader may object: Which is it when teachers in Reasoning Mind classrooms walk around the room, answering student questions? These questions are sometimes about the content (e.g., "how do you solve this problem?") and other times more general (e.g., "may I please use the bathroom?"). In this case, I would say the teacher is mainly engaged in facilitation, but periodically shifts into instruction (when answering content questions) for brief periods of time. In other words, the facilitation/instruction distinction is meant to be applied to each moment rather than to broad activities lasting many minutes.

skills: teachers must establish enough structure, trust, and consistency that students behave as required even in the absence of the watchful eye. This is not an impossible task, but nor is it a simple one: a minority of teachers—albeit a sizable one—succeed completely, but most achieve only a partial victory.

There is a second, more subtle problem. People are not good at distinguishing between activity and productivity. For example, shortly after obtaining a cell phone that could receive e-mail, I developed a habit of checking it every several minutes. I realized it was a nervous tic, likely annoying to those who shared my company, but did not think it had further consequences. When, years later, I finally eradicated the habit, I found to my surprise that my productivity in tasks that required extended focus (such as writing this book) increased nearly twofold. For all those years, I had thought that I was productively occupied, going about my work while staying cleverly apprised of the latest e-mails; in reality, I had been wasting half my time. This sad story carries a broader lesson: when we are occupied, we often assume we are being productive. The assumption is often wrong.* In the blended learning classroom, this phenomenon has unfortunate consequences. A teacher who spends her time exclusively on facilitation—walking around the room, monitoring students, answering questions, and offering encouragement—feels herself to be productively occupied. She will end her day with a sense of satisfaction at a job well done. Thus, the fact that facilitation by itself is enough to *occupy* teachers leads to an insufficient allotment of time to instruction. I have seen this countless times in my visits to blended learning classrooms: the best teachers will spend nearly all of their time engaged in instruction, but the majority will spend only half or less. This carries a corollary waste: highly trained teachers, whose charge is to know mathematics and to expand its empire, instead spend much of their time on administrative activities—which require energy, but considerably less professional training.

For some blended learning programs—for example, those that see the computer as the central source of instruction—this is not a problem. But in instruction modeling, the aim is to provide the

* An economist might say that we overlook the *opportunity cost* associated with how we spend our time. Ignoring opportunity costs is a well-known bias, found in a wide range of situations.

full range of experiences students receive in exceptional classrooms, and this includes those experiences that cannot be automated. If the classroom teacher does not devote a great deal of time to instruction, we cannot say that we have an accurate model.

What can be done? In the short term, teachers can be taught the importance of instructional engagement and encouraged to make it a priority. Indeed, Reasoning Mind's implementation staff expend considerable effort to this end. But in the long run, the most natural solution is to split the two roles. There should be separate facilitators—who are not expected to provide any instruction—and teachers, who are free to focus on instruction alone. This would be a decisive resolution.

This split would require a wholesale redesign of the school. Instead of rooms made to fit twenty or thirty students, with schedules meant for lockstep progression, the school would need separate rooms for computer lessons and for meetings with teachers. For computer lessons, as many as sixty students* could work in a large room, supervised by a single facilitator. Meanwhile, teachers would meet with individual students, small groups, or seminars of a dozen students in separate rooms appropriate to these formats. This would ensure that facilitation and instruction each receive their proper due.

This approach is not without a venerable precedent. In the nineteenth century, Benjamin Jowett, master of Balliol College at Oxford and subsequent vice-chancellor of the university, championed the adoption of what later became known as the *Oxford tutorial method*.[12] In this method of instruction, besides attending large-group lectures, students meet weekly for "tutorials" with professors. A tutorial might include one, two, or at most three students. Students arrive with a completed work product: in history, an essay; in mathematics, a problem set. Each student presents his work, reading aloud to the professor. She listens carefully and questions the student, probing his reasoning.** Modeled on a Socratic dialogue—Jowett was an expert in Plato—the

* Lest you think this overly optimistic, IDEA Public Schools has successfully operated computer labs with a single facilitator watching over dozens of students.

** This might remind you of "work at the board," an activity widely practiced by Russian mathematics teachers. It was discussed in chapter 5.

tutorial offers exactly the kind of experience that cannot be given in a large class, and certainly cannot be provided by a computer lesson.

The Oxford tutorial method subsequently spread to Cambridge, and to dozens of other universities around the world. In the United States, Williams College offers tutorial method instruction to seniors, and Ohio University has an honors college that uses the method exclusively. However, it is apparent why the method did not take similar hold in schools: lectures are unsuitable for school-age children, who have not yet developed the independent learning abilities of undergraduates. School students need more structure, such as that provided by an interactive online lesson. Thus, blended learning finally affords the possibility of extending the Oxford tutorial method to schools. Doing so would conclusively resolve the tension between facilitation and instruction in implementations of blended learning, and would yield the ideal environment for the use of programs developed by means of instruction modeling.

Of course, creating such schools would be a large undertaking. Moreover, there is a curious dilemma: such a school can only succeed if comprehensive blended learning programs are available for a wide range of school subjects, but such programs are only likely to be developed if there is demand for them from schools. This is one reason why technology-based reform is challenging: programs and schools must all be redesigned in concert, an immense undertaking.

Fortunately, there is increased awareness of the need for such comprehensive redesign. For example, a new organization called Transcend, founded by Aylon Samouha and Jeff Wetzler, advocates for the development of fundamentally new school models, along with the technologies they require. It is entirely possible that twenty years from now, schools will be built and operated according to completely different principles from those of today. If the programs these schools use are instructionally sound, then this will be to the great benefit of the millions of students who will attend them.

Government Policy

In chapter 5, I discussed the unfortunate circumstance that comprehensive blended learning programs (such as programs developed using instruction modeling) necessarily spread slower, even though they are more instructionally effective. One reason for this is a misalignment of incentives: schools must focus on increasing (or at any rate maintaining) short-term scores on high-stakes tests, and for this purpose the degree of alignment between their programs and the tests far outweighs any benefits of increased instructional quality. Comprehensive programs, with all of the difficulties inherent in their use, are in this sense a luxury: they matter educationally, but make no discernible difference administratively.

Tests are the culprit: they offer a good general sense of what students know, but are a more limited measure of how well students were taught in any given year. Thus, when test scores are used as a proxy for instructional quality, the former soar while the latter languishes.[13]

Does this mean testing should be abolished? There are many who think so, just as there are many who think the opposite. But the picture, as is usual in education, is complicated. For example, Stanford historian of education Larry Cuban—a brilliant and nuanced thinker on educational questions—notes that testing has consistently exposed educational inequities, such as disparities between the rich and poor, and between minorities and majorities, at the same schools.[14] And if testing is altogether abolished, what measures will remain in place to guide the fragmented school systems of each state toward a common educational standard? And how will parents and state governments be able to identify those schools that are failing and in need of intervention?

Comprehensive blended learning programs offer an exciting solution to this dilemma. Unlike a test, which offers a snapshot at one point in time, these programs see students every day. They know what topics students have mastered, and what topics they struggle with; they know when students have gone beyond the minimum requirements, solving challenging advanced problems; they know when students have done their homework. These programs do not need tests to provide additional information: they *are* tests, and far more complete ones than anything that can be taken in a day. Valerie Shute, a leading psychometrician at Florida State University, calls this "stealth assessment": an online program, just like a teacher, can form an accurate picture of what students know without the need for anything resembling an explicit exam.[15] Mingyu Feng and her colleagues have shown that this kind of assessment can very accurately match standardized test scores in practice.[16] Instruction and assessment are one and the same. In some sense, this is nothing new: college courses are traditionally graded based on a combination of scores; the final exam counts, but so does homework, midterm exams, quizzes, essays, and even class participation. Comprehensive blended learning programs make it possible to bring the same to school accountability systems, either replacing or—more likely—supplementing end-of-year exams. You might call it "comprehensive blended assessment."

States could change their accountability systems to implement such an approach. A state would review blended learning programs—just as textbooks are reviewed and approved by many states today—and select the ones that are suitable. Then, schools could choose whether to continue with the conventional accountability system—under which the end-of-year test is the sole measure—or to implement an approved blended learning program and enroll in the new, blended accountability system. For these participating schools, the end-of-year test would still make up some portion (for example, a third) of their accountability results, with the remainder generated by the blended learning program based on students' performance throughout the year.

What would the benefits be of such a policy? In my view, there are many. First, it would yield a more accurate assessment of students' performance, since it would rely on a much richer and more complete dataset. A student's score on a test depends on how she was feeling that morning, whether she happened through luck to have seen very similar questions previously, whether she slept well, whether she guessed successfully for questions she did not know, how nervous she was, and many other such things. The new system greatly reduces these distortions.

Second, the new system would reduce the time spent on test preparation and practice testing. Today, schools devote inordinate time to these activities, partly to prepare students for the tests and partly to ensure there are no great surprises when the final results come in.[17] The new system would still call for some test preparation, but to a substantially reduced extent—freeing more time for instruction.

Third, the system would provide continuous information on where students stand. Instead of waiting until the end of the year to find out, students, teachers, and parents could see beforehand how students were doing on their accountability ratings, and intervene accordingly. Of course, this would greatly relieve the stress of teachers, school administrators, parents, and children; today, too many third graders spend their spring evenings anxious about their upcoming tests. But it would also offer better-quality motivation, since students would see an immediate link between their actions (working hard in mathematics class) and the reward (good accountability scores).*

Fourth, it would encourage strong fidelity of implementation for the blended learning programs. Instead of time online being in conflict with time spent on test preparation, participating schools would see it as a key to success in the accountability system. Teachers would see the immediate benefits of tutoring students who struggle with online topics, since these students' success would be directly linked to their scores. And every important topic or idea covered online—regardless of whether it is highly likely to show up as a multiple-choice problem on a test—would be seen as important. In other words, there would be a much tighter link between the accountability results and the daily demands of good instruction. Instead of clashing with the fidelity of implementation, the accountability system would support it.

And fifth, it would encourage school districts to adopt core curriculum blended learning programs. Today, schools balk at such radical changes for fear of damaging their standing in accountability: any large change carries the potential of failure, and schools are understandably averse to such risks. This leads schools to adopt trifling programs, meant to be used lightly as a diversion, over the more substantial fare that has the potential to meaningfully improve instruction. But if the adoption of substantial programs would give schools the opportunity to obtain accountability scores gradually over the course of the year—rather than in a single fell swoop at the end—it would provide the opposite: a strong motive for trying the new approaches. In other words, the new system would reward innovation instead of punishing it.

* For example, in one study (Fryer, 2010), some children were paid for receiving higher grades in English classes while others were paid for reading books. The first group saw no improvement, while the second both read more and improved their English grades. When rewards are more directly linked to actions, people see a clear path to success and are more likely to make a constructive effort.

Of course, there are also some significant pitfalls that must be avoided. To begin with, only comprehensive programs—meant as core curricula or intensive (fifty hours annually or more) ancillary ones—could be included. This is because the entire premise of this model is to embed assessment in regular *instruction;* if a program is only used sparingly, it becomes merely another assessment, and many of the aforementioned benefits vanish. This is a limitation of the proposed policy, since today there are few comprehensive blended learning programs to choose from. However, the policy would encourage the creation of more, since the paucity of comprehensive programs is a consequence of the limited demand for such programs on the part of schools—a circumstance that the proposed policy would reverse.

Another pitfall concerns program quality. If a program is instructionally weak, then allowing schools to use it in this manner would be a disservice: students and teachers would work hard to successfully implement a program that is not worth implementing at all. This means states must carefully vet programs, choosing those which not only cover appropriate material, but also—and far more importantly—do so in a way that is mathematically and instructionally sound. The result will only be as good as the judgment of states in deciding which programs to adopt.

Finally, there is always the concern of cheating. Schools can cheat even on end-of-year exams—and many do—but this system brings accountability into the less stringently regulated environment of the regular classroom. Would it cause fraud on a massive scale? In my view, this is unlikely. The same honor code that prevents the vast majority of teachers from cheating on end-of-year exams would prevent the behavior here; most people, after all, are honest. Furthermore, cheating would in some ways be *more* difficult than on the tests: it is one thing to change the multiple choice options marked by students on a form, but another thing altogether to ensure that thirty students working on thirty completely different problems all are fed the correct answer. No teacher, no matter how brilliant an arch-villain, could solve problems at this pace. Finally, if some cheating did end up taking place, it would be easy to build statistical "detectors" (of the sort discussed in chapter 7 and then again earlier in the present chapter) to identify likely occurrences.[18] The Securities and Exchange Commission uses such methods to spot insider trading, and I see no reason why insider testing should not receive the same treatment.

While these pitfalls all seem readily avoidable, wise states will begin by *piloting* the new approach. For example, a state might create a so-called "innovation zone," a consortium of schools committed to trying the new approach for several years, learning from the experience and perfecting the system to prepare it for general use.

In my view, government policy has an essential role to play in making the widespread use of effective blended learning possible. By realigning accountability systems with program implementation quality, states can obtain more accurate data; reduce the stress of children, parents, and educators; move the focus to instruction rather than superficial test preparation; encourage schools to use high-quality blended learning programs; and encourage publishers to develop comprehensive programs that carry the potential for meaningful instructional change. State governments can be the unmoved movers, setting in motion a cascade of virtuous events. With the right government policies, core curriculum blended learning can become the norm rather than the exception in our nation's classrooms.

Conclusion
What of It?

WHAT ARE THE LESSONS OF THIS BOOK? PERHAPS THE most general lesson is something that might appear to be common sense: *a blended learning program can only fundamentally improve instruction if it is comprehensive*—i.e., if it addresses *all* of the aspects of good instruction in the subject, and not just some. In particular, a program cannot be comprehensive if it does not take the appropriate *time*: programs used for a few hours a year can be nothing more than diversions, and those used for a few dozen hours can be helpful but not transformative. For a program to make the difference between middling and excellent instruction, it must be a core curriculum. This does not mean that anything besides core curriculum is worthless: such programs can be valuable intermediate steps. But we must remember the end goal, and we should never fall into the trap of thinking that a program can have an effect out of proportion with its actual usage. You cannot grow an oak tree in a year, no matter how attractively the acorn was packaged at the store.

Beyond this, we should remember that not all blended learning programs are equal in quality. They are not commodities. Programs are less like paperclips and more like films. Some are good, and some

Instruction Modeling. George A. Khachatryan, Oxford University Press (2020)

DOI: 10.1093/oso/9780190910709.001.0001

are frightful. Not everyone can tell the difference. Determining the instructional worth of a program is a challenging and nuanced undertaking: at a minimum, it requires a broad knowledge of education research, multiple field experiments, and a detailed study of the program by genuine experts in the discipline. A striking graph shown by a salesperson is a poor substitute for these things.

What determines the ultimate instructional success of a blended learning program? There are three dimensions of quality: content, interface, and implementation. The content is the substance of the instruction. The interface is the particular way that substance is conveyed. The implementation is the way the program is used in practice: it determines whether the *experienced* content and interface end up corresponding to the *intended* ones. As an example, consider a program consisting of static text webpages describing ways to add and subtract fractions; the content is the substance of what is written, the interface is static text (meant to be read by students), and the implementation is the way the pages are actually used in the classroom.

For a program to be instructionally strong, each dimension must be strong. The content should offer all of the ideas and experiences needed for mastery of the material; the interface should be of the kind that best encourages active thought and comprehension; and as for implementation, the program should include the training and support needed to allow teachers to use it as intended.

All three dimensions are to varying degrees neglected in the design of blended learning programs today. The most neglected one is content. Partly this is because of the difficulty in telling apart good content from bad. And perhaps program designers underestimate the great difficulty of creating good content: in the world's best instructional traditions, the content (which includes not only textbooks, but also the details of each lesson delivered by teachers) was perfected over decades. Excellent content does not merely conform with a set of general principles—though this, too, is true—but rather is outstanding in the details of each moment. This requires of teachers and curriculum developers deep knowledge of mathematics and of children, but even then can only be achieved through constant, deliberate trial and error over many years. And furthermore, good content must be coherent: it is not a disjointed collection of pieces, each "good" individually, but rather a connected progression, with the individual components fitting together into a logical whole. Given all this, it is no wonder that good content is so rare.

The difficulty of creating good content is the primary motivation for instruction modeling. Instead of attempting to create it from scratch, the instruction modeler seeks good content where it already exists: outstanding traditions of instruction. Because the quality of content resides in the details, instruction modeling is necessarily a laborious process. In the ideal, every minute of every lesson should be studied and carefully translated to the new medium. This kind of meticulous labor is inherent in instruction modeling, and—more generally—I am not aware of any consistent method to create good content that avoids it.

Beyond content, instruction modeling offers some help with the other two dimensions, interface and implementation. However, it does not provide here the same complete guidance it offers in the development of content. For this reason, the astute designer uses instruction modeling in concert with other approaches, chief among them the systematic application of learning science.

The evidence—as we discussed in chapter 6—indicates that programs developed using instruction modeling succeed in their aims. Students using these programs learn more and better mathematics, spend more time on task and engaged, and report enjoying the experience. Teachers and administrators like the programs and find that they help their students. And implementation fidelity is strong: in core curriculum programs, students average over a hundred hours online annually, and the extensive training and implementation support help teachers understand and use the program as intended.

However, as chapter 9—in which we discussed the likely future of instruction modeling—made clear, a long road remains before instruction modeling can achieve its full potential. Many exciting improvements can be made, from incorporating new technologies (such as those afforded by natural language processing and by big data in education), to expanding to subjects beyond mathematics, to designing schools that are specially structured to take full advantage of comprehensive blended learning programs. Each of these improvements will be a large undertaking, no doubt full of many surprises and important lessons.

But beyond improving the practice of instruction modeling, a second great challenge remains: expanding the use of programs developed with its aid. This problem is not unique to programs developed using instruction modeling, but rather concerns *any* comprehensive blended learning program. The current system of incentives for schools discourages the adoption of comprehensive programs—even though it is comprehensive programs alone that offer the possibility of meaningful improvement in instruction. The only decisive solution is the creation of improved government policies, which can better align the pressures on schools with the actions needed for instructional excellence.

What can each of us do to ensure that instruction modeling achieves its full potential? I believe there are two things above all else. First, we must realize that instruction modeling—and blended learning in general—is still taking its very first steps. A great deal of further research and development will be needed before it enters its full maturity. We must continue learning, experimenting, and pushing the boundaries of what can be accomplished.

And second, we must educate ourselves and the public: our colleagues, friends, and innocent bystanders. Education is a complex, subtle field. There are no simple solutions. Separating the wheat from the chaff is hard. Because of this, we cannot assume that the best approaches—like the best mobile phone models, or the best restaurants in a crowded city—will rise to the top of their own accord. Technology offers great possibilities in education, but its potential can be easily squandered; bad instruction can be technologized as easily as good. Blended learning is like a radio: it can send music far and wide, but the music we choose to send is up to us. For this reason, the quality of programs used in our schools will always and forever be directly tied to our collective discernment. By constantly learning more, digging deeper, and sharing our insights with one another, we can develop ever-stronger judgment in educational matters, to the benefit of our societies and the millions of children in our schools.

The effort is considerable, but the rewards are great. If we succeed, we can dramatically improve the instruction offered in our schools. Teachers will be freed of clerical tasks, offered the professional development needed for mastery of their craft, and given the chance to focus their full energies on instruction. Students across the country and the world will have the same exceptional instructional experiences that today are the province of a fortunate few. Countless children will discover the pleasure of thinking, the delight of understanding, and the great joy of the use of those remarkable faculties that make us distinctly human.

I had two great strokes of mathematical luck in my childhood: my grandmother's intervention and the chance to attend Mathcamp. These experiences changed my life. Instruction modeling offers the prospect of bringing the same opportunities to many more children.

Notes

Introduction

1. Buchanan and Shortliffe, 1984.

Chapter 1

1. The first view was held by Plato (cf., *The Republic*), the second argued by historian Jacques Barzun (1989) in his essay "Where the Educational Nonsense Comes From."
2. Ravitch, 2001.
3. Rohlen and Bjork, 1998.
4. Lindberg, 2010.
5. Hergenhahn and Henley, 2013.
6. Peirce and Jastrow, 1884; Hall, 2007.
7. Thomas et al., 2016.
8. Clark and Mayer, 2016.
9. Open Science Collaboration, 2015.
10. See Nystrand and Gamoran (1991) for the importance of authentic discussion in literature classes, and Clark and Mayer (2016) for the other points.
11. Cameron, 1963.

12. Barzun, 1989. Essay: "Culture, High and Dry."
13. Hardy, 1992. The essay was first published in 1940.
14. Koretz, 2008, page 9.
15. Koretz, 2008, page 245. The original article reporting this result is Koretz et al., 1991.
16. One salient example is a study of students in KIPP schools (Tuttle et al., 2010). The meticulously designed study found that KIPP middle school students improved their scores by roughly one third of a grade's worth in each year. This is a substantial amount, and quite reasonable to expect given the schools' approach: rigorous discipline, increased learning time on weekdays and weekends, a focus on tested subjects to the exclusion of curricular breadth, and exceptionally motivated teachers.
17. Gomes, "Five Things About Efficacy That Should Be Intuitive, But Are Not," retrieved from www.edsurge.com.
18. Koretz, 2008.
19. Cuban, 1986.
20. Ma, 1999.
21. Barzun, 1989.
22. Feynman, 2010.
23. Stigler and Hiebert, 2009.
24. Ibid.
25. Ibid.

Chapter 2

1. Ravitch, 2001.
2. Ibid.
3. Schoenfeld, 2004.
4. Mayer, 2004.
5. Chi et al, 1994.
6. For example, some studies have found that it can be beneficial to allow students to explore a problem in an unguided manner first and provide direct instruction afterwards. See DeCaro and Rittle-Johnson, 2012 as well as Schwartz and Martin, 2004.
7. Stigler and Hiebert, 2009.
8. Ibid.
9. Ibid.
10. Jones, 1977.

11. Parker, 2005.
12. Ferry, 2006.
13. Willingham, 2009. Willingham takes as his central point in the chapter of remembering: "memory is the residue of thought."
14. Gick and Holyoak, 1983.
15. Chi et al., 1981.
16. Schmidt et al., 2002.
17. Willingham, 2009.
18. Wu, 1999.
19. Willingham, 2009.
20. Rittle-Johnson et al., 2015. The authors argue that concepts and procedures support one another and survey the evidence on the subject. Many researchers have argued in favor of providing conceptual instruction before procedural instruction, but evidence directly demonstrating the superiority of this approach is scarce.
21. Ravitch, 2001.
22. Gick and Holyoak, 1983.
23. Willingham, 2009.
24. Plomin and Deary, 2015.
25. Brinch and Galloway, 2012.
26. Veenman and Alexander, 2011.
27. Polya, 2014.
28. Givental, n.d.
29. Karp and Vogeli, 2010.
30. Stigler and Hiebert, 2009.
31. Krug, 2005.
32. Schmidt et al., 2002.
33. Givental, n.d.
34. Stigler and Hiebert, 2009.
35. Tucker, 2011.
36. Hill et al., 2005.
37. Institute of Education Sciences, 2013.
38. Stigler and Hiebert, 2009.
39. Ibid.
40. Ma, 1999.
41. Givental, n.d.
42. Wu, 2011.
43. Tucker, 2011.

44. Darling-Hammond, 2010.
45. Stigler and Hiebert, 2009.
46. Koretz, 2008.

Chapter 3

1. Milgram, 2005.
2. Karp and Vogeli, 2010.
3. Stigler and Hiebert, 2009.
4. Givental, n.d.
5. Stigler and Hiebert, 2009.
6. Jackson, 1986.
7. Russell et al., 2003.
8. Buchanan and Shortliffe, 1984.
9. Ibid.
10. Russell et al., 2003.
11. Clancey, 1987.
12. Jackson, 1986.
13. Stigler and Hiebert, 2009.
14. Koretz, 2008.
15. Karp and Vogeli, 2010.
16. Ibid.
17. Milgram, 2005.
18. Lando, 2015.
19. Khachatryan et al., 2014.
20. Lando, 2015.
21. Weber, 2003.
22. Buchanan and Shortliffe, 1984.

Chapter 4

1. Clark and Mayer, 2016.
2. Ibid.
3. Ibid.
4. Ibid.
5. Ibid.

Chapter 5

1. Ibid.
2. Studies have found that "engaged concentration" (which is when students are quietly focused on a learning task) is predictive of positive learning outcomes. See, for example, Pardos et al., 2013.
3. Means and Peters, 2016.

Chapter 6

1. Clark and Mayer, 2016.
2. Cuban, 1986.
3. Horn and Staker, 2014.
4. Hess and Saxberg, 2013.
5. The textbooks were primarily authored by Pittsburgh public school teacher Bill Hadley. See, for example, the timeline of the development of cognitive tutors on the Cognitive Tutor Authoring Tools (CTAT) site at Carnegie Mellon University. Retrieved from http://ctat.pact.cs.cmu.edu/.
6. Posovac, 2015.
7. Carrell and West, 2008; Braga et al., 2011.
8. Baseline numbers can be found in studies of engagement in conventional classrooms, such as Lee, Kelly, & Nyre, 1999. Studies reporting engagement measurements in blended learning classrooms include Pardos et al., 2013 as well as Baker et al., 2012.

Chapter 7

1. Ma, 1999.
2. Stigler and Hiebert, 2009.
3. Hill et al., 2005.
4. Hess and Saxberg, 2013.
5. Smith, 1984.
6. Willingham (2009) devotes considerable attention to debunking the myth of learning styles.

7. In Twain's defense, he also spoke frequently in support of education.
8. Bloom, 1984.
9. VanLehn, 2011.
10. Hattie and Gan, 2011.
11. Willingham, 2009.
12. Walkington, 2013.
13. Corbett and Anderson, 1994; Doignon and Falmagne, 1985.
14. Bulmer, 2012.
15. James et al., 2013.
16. Romero and Ventura, 2007.
17. Baker et al., 2004.
18. Baker et al., 2010.

Chapter 8

1. Heffernan &and Koedinger, 2002.
2. Graesser et al, 2005.
3. Haidt, 2012.
4. Stigler & Hiebert, 2009.
5. Buchanan and& Shortliffe, 1984.
6. Willingham, 2009.
7. For example, Koedinger et al., 1997.
8. VanLehn, 2011.

Chapter 9

1. Chi et al., 1994.
2. Chi and Wylie, 2014.
3. Fonseca and Chi, 2011.
4. McNamara et al., 2004.
5. Graesser et al., 2005.
6. Slavin, 1990.
7. McLaren et al., 2005.
8. Nystrand and Gamoran, 1991.
9. Roschelle et al., 2016.
10. Baker et al., 2012.
11. Miller et al., 2015.

12. Ashwin, 2005.
13. Koretz, 2008.
14. David and Cuban, 2010.
15. Shute and Ventura, 2013.
16. Feng et al., 2009.
17. Nichols et al., 2007.
18. Jacob and Levitt, 2003.

References

Ashwin, P. (2005). Variation in students' experiences of the' "Oxford tutorial'." *Higher Education, 50*(4), 631–644.

Baker, R. S., Corbett, A. T., Koedinger, K. R., & Wagner, A. Z. (2004, April). Off-task behavior in the cognitive tutor classroom: When students game the system. In *Proceedings of the SIGCHI conference on human factors in computing systems* (pp. 383–390). ACM.

Baker, R. S., D'Mello, S. K., Rodrigo, M. M. T., & Graesser, A. C. (2010). Better to be frustrated than bored: The incidence, persistence, and impact of learners' cognitive–affective states during interactions with three different computer-based learning environments. *International Journal of Human-Computer Studies, 68*(4), 223–241.

Baker, R. S., Gowda, S. M., Wixon, M., Kalka, J., Wagner, A. Z., Salvi, A., . . . Rossi, L. (2012). Towards sensor-free affect detection in cognitive tutor algebra. In *Proceedings of the 5th international conference on educational data mining* (pp. 126–133).

Barzun, J. (1989). *The culture we deserve*. Middletown, CT: Wesleyan University Press.

Bloom, B. S. (1984). The 2 sigma problem: The search for methods of group instruction as effective as one-to-one tutoring. *Educational Researcher, 13*(6), 4–16.

Boriack, A. W., Stillisano, J. R., Wright, K. B., Waxman, H. C. (2015). Independent evaluation of Reasoning Mind programming. Retrieved from http://www.reasoningmind.org.

Braga, M., Paccagnella, M., & Pellizzari, M. (2014). Evaluating students' evaluations of professors. *Economics of Education Review, 41*, 71–88.

Brinch, C. N., & Galloway, T. A. (2012). Schooling in adolescence raises IQ scores. *Proceedings of the National Academy of Sciences, 109*(2), 425–430.

Buchanan, B. G., & Shortliffe, E. H. (Eds.). (1984). *Rule-based expert systems* (Vol. 3). Addison-Wesley.

Bulmer, M. G. (2012). *Principles of statistics*. Courier Corporation.

Cameron, W. B. (1963). *Informal sociology: A casual introduction to sociological thinking* (Vol. 21). Random House.

Carrell, S. E., & West, J. E. (2008). *Does professor quality matter? Evidence from random assignment of students to professors* (No. w14081). National Bureau of Economic Research.

Chi, M. T., De Leeuw, N., Chiu, M. H., & LaVancher, C. (1994). Eliciting self-explanations improves understanding. *Cognitive Science, 18*(3), 439–477.

Chi, M. T., Feltovich, P. J., & Glaser, R. (1981). Categorization and representation of physics problems by experts and novices. *Cognitive science, 5*(2), 121–152.

Chi, M. T., & Wylie, R. (2014). The ICAP framework: Linking cognitive engagement to active learning outcomes. *Educational Psychologist, 49*(4), 219–243.

Clancey, W. J. (1988). *Knowledge-based tutoring: The GUIDON program*. MIT Press Series in Artificial Intelligence.

Clark, R. C., & Mayer, R. E. (2016). *E-learning and the science of instruction: Proven guidelines for consumers and designers of multimedia learning*. Wiley.

Corbett, A. T., & Anderson, J. R. (1994). Knowledge tracing: Modeling the acquisition of procedural knowledge. *User Modeling and User-Adapted Interaction, 4*(4), 253–278.

Cuban, L. (1986). *Teachers and machines: The classroom use of technology since 1920*. Teachers College Press.

Darling-Hammond, L. (2010). Teacher education and the American future. *Journal of Teacher Education, 61*(1–2), 35–47.

Darling-Hammond, L. (2015). *The flat world and education: How America's commitment to equity will determine our future.* Teachers College Press.

David, J. L., & Cuban, L. (2010). *Cutting through the hype: The essential guide to school reform. Revised, expanded.* Harvard Education Press.

DeCaro, M. S., & Rittle-Johnson, B. (2012). Exploring mathematics problems prepares children to learn from instruction. *Journal of Experimental Child Psychology, 113*(4), 552–568.

Dewey, J. (1986, September). Experience and education. In *The Educational Forum* (Vol. 50, No. 3, pp. 241–252). Taylor & Francis Group.

Doignon, J. P., & Falmagne, J. C. (1985). Spaces for the assessment of knowledge. *International Journal of Man-Machine Studies, 23*(2), 175–196.

Douglas, D., Gaudino, S. J., Khachatryan, G. A., Kostyuk, V. A., & Yufa, N. (2012). Report on the 2011–2012 Reasoning Mind implementation at Dallas ISD. Retrieved from http://www.reasoningmind.org.

Feng, M., Heffernan, N., & Koedinger, K. (2009). Addressing the assessment challenge with an online system that tutors as it assesses. *User Modeling and User-Adapted Interaction, 19*(3), 243–266.

Ferry, D. (2006). *The Georgics of Virgil: Bilingual Edition.* Macmillan.

Feynman, R. P. (2010). *"Surely 'You're Joking, Mr. Feynman!": Adventures of a Curious Character.* Norton & Company.

Fonseca, B. A., & Chi, M. T. (2011). Instruction based on self-explanation. *Handbook of Research on Learning and Instruction,* 296–321.

Fryer, R. G., Jr. (2010). Financial incentives and student achievement: Evidence from randomized trials (No. w15898). National Bureau of Economic Research.

Gick, M. L., & Holyoak, K. J. (1983). Schema induction and analogical transfer. *Cognitive Psychology, 15*(1), 1–38.

Givental, A.B. (n.d.). How the incompetent enlightens the ignorant. Retrieved from: https://math.berkeley.edu/~giventh/.

Graesser, A. C., Chipman, P., Haynes, B. C., & Olney, A. (2005). AutoTutor: An intelligent tutoring system with mixed-initiative dialogue. *IEEE Transactions on Education, 48*(4), 612–618.

Haidt, J. (2012). *The righteous mind: Why good people are divided by politics and religion*. Knopf Doubleday Publishing Group.

Hall, N. S. (2007). RA Fisher and his advocacy of randomization. *Journal of the History of Biology, 40*(2), 295–325.

Hardy, G. H. (1992). *A mathematician's apology*. Cambridge University Press.

Hattie, J., & Gan, M. (2011). Instruction based on feedback. In R. E. Mayer, & P. A. Alexander (Eds.), *Handbook of research on learning and instruction* (pp. 249–271). New York: Routledge.

Heffernan, N. T., & Koedinger, K. R. (2002). An intelligent tutoring system incorporating a model of an experienced human tutor. *International Conference on Intelligent Tutoring Systems*, 596–608.

Hergenhahn, B. R., & Henley, T. (2013). *An introduction to the history of psychology*. Cengage Learning.

Hess, F. M., & Saxberg, B. V. H. (2013). *Breakthrough leadership in the digital age: Using learning science to reboot schooling*. Corwin.

Hill, H. C., Rowan, B., & Ball, D. L. (2005). Effects of teachers' mathematical knowledge for teaching on student achievement. *American Educational Research Journal, 42*(2), 371–406.

Horn, M. B., & Staker, H. (2014). *Blended: Using disruptive innovation to improve schools*. Jossey-Bass.

Institute of Education Sciences. (2013). U.S. states in a global context: Results from the 2011 NAEP-TIMSS linking study. Retrieved from https://nces.ed.gov.

Jackson, P. (1986). *Introduction to expert systems*. Addison-Wesley.

Jacob, B. A., & Levitt, S. D. (2003). Rotten apples: An investigation of the prevalence and predictors of teacher cheating. *Quarterly Journal of Economics, 118*(3), 843–877.

James, G., Witten, D., Hastie, T., & Tibshirani, R. (2013). *An introduction to statistical learning* (Vol. 6). New York: Springer.

Jones, F. B. (1977). The Moore method. *American Mathematical Monthly, 72*, 273–278.

Karp, A., & Vogeli, B. R. (Eds.). (2010). *Russian mathematics education: History and world significance*. World Scientific.

Khachatryan, G. A., Romashov, A. V., Khachatryan, A. R., Gaudino, S. J., Khachatryan, J. M., Guarian, K. R., & Yufa, N. V. (2014). Reasoning Mind Genie 2: An intelligent tutoring system as a vehicle for international transfer of instructional methods in mathematics. *International Journal of Artificial Intelligence in Education, 24*(3), 333–382.

Koedinger, K. R., Anderson, J. R., Hadley, W. H., & Mark, M. A. (1997). Intelligent tutoring goes to school in the big city. *International Journal of Artificial Intelligence in Education (IJAIED), 8*, 30–43.

Koretz, D. M. (2008). *Measuring up*. Harvard University Press.

Koretz, D., Linn, R. L., Dunbar, S. B., & Shepard, L. A. (1991). The effects of high-stakes testing: Preliminary evidence about generalization across tests. In *The Effects of High Stakes Testing*, ed. R. L. Linn (chair). Symposium presented at the annual meetings of the American Educational Research Association and the National Council on Measurement in Education. Chicago, April.

Krug, S. (2005). *'Don't make me think: A common sense approach to web usability*. Pearson Education India.

Lando, S. (2015). Mathematical education in universities in the Soviet Union and modern Russia. *Higher Education in Russia and Beyond, 4*(6), 6–8.

Lee, S. W., Kelly, K. E., & Nyre, J. E. (1999). Preliminary report on the relation of students' on-task behavior with completion of school work. *Psychological Reports, 84*(1), 267–272.

Lindberg, D. C. (2010). *The beginnings of Western science: The European scientific tradition in philosophical, religious, and institutional context, prehistory to AD 1450*. University of Chicago Press.

Ma, L. (1999). *Knowing and teaching elementary mathematics: 'Teachers' understanding of fundamental mathematics in China and the United States*. Lawrence Erlbaum Associates.

Mayer, R. E. (2004). Should there be a three-strikes rule against pure discovery learning? *American Psychologist, 59*(1), 14.

McLaren, B. M., Bollen, L., Walker, E., Harrer, A., & Sewall, J. (2005, May). Cognitive tutoring of collaboration: Developmental and empirical steps towards realization. In *Proceedings of the 2005 conference on Computer support for collaborative learning: Learning 2005: the next 10 years!* (pp. 418–422). International Society of the Learning Sciences.

McNamara, D. S., Levinstein, I. B., & Boonthum, C. (2004). iSTART: Interactive strategy training for active reading and thinking. *Behavior Research Methods, Instruments, and Computers, 36*(2), 222–233.

Means, B., & Peters, V. (2016). Influences on the scaling of digital learning resources. Paper presented at the AERA symposium "Scaling Up Digital Media Innovations."

Milgram, R. J. (2005). The mathematics pre-service teachers need to know. Retrieved from ftp://math.stanford.edu/pub/papers/milgram/FIE-book.pdf.

Miller, W. L., Baker, R. S., Labrum, M. J., Petsche, K., Liu, Y. H., & Wagner, A. Z. (2015, March). Automated detection of proactive remediation by teachers in Reasoning Mind classrooms. In *Proceedings of the Fifth International Conference on Learning Analytics and Knowledge* (pp. 290–294). ACM.

Mingle, L., Khachatryan, G., and Kostyuk, V. (2017a). Adapting international math curricula using an online platform. Poster presented at AERA 2017 Annual Meeting.

Mingle, L., Khachatryan, G., and Kostyuk, V. (2017b). Evaluating the Reasoning Mind fifth-grade common core curriculum. Paper presented at AERA 2017 Annual Meeting.

Mulqueeny, K., Kostyuk, V., Baker, R. S., & Ocumpaugh, J. (2015). Incorporating effective e-learning principles to improve student engagement in middle-school mathematics. *International Journal of STEM Education*, *2*(1), 1.

Nichols, S. L., Berliner, D. C., & Noddings, N. (2007). Collateral damage: How high-stakes testing corrupts America's schools. Harvard Education Press, 2007.

Nystrand, M., & Gamoran, A. (1991). Instructional discourse, student engagement, and literature achievement. *Research in the Teaching of English*, *25*(3), 261–290.

Ocumpaugh, J., d Baker, R. S., Gaudino, S., Labrum, M. J., & Dezendorf, T. (2013, July). Field observations of engagement in Reasoning Mind. In *International Conference on Artificial Intelligence in Education* (pp. 624–627). Springer.

Open Science Collaboration. (2015). Estimating the reproducibility of psychological science. *Science*, *349*(6251), aac4716.

Pardos, Z. A., Baker, R. S., San Pedro, M. O., Gowda, S. M., & Gowda, S. M. (2013, April). Affective states and state tests: Investigating how affect throughout the school year predicts end of year learning outcomes. In *Proceedings of the Third International Conference on Learning Analytics and Knowledge* (pp. 117–124). ACM.

Parker, J. (2005). *RL Moore: Mathematician and teacher*. MAA.

Peirce, C. S., & Jastrow, J. (1884). On small differences in sensation. *Memoirs of the National Academy of Science, 3*, 75–83.

Plomin, R., & Deary, I. J. (2015). Genetics and intelligence differences: Five special findings. *Molecular Psychiatry, 20*(1), 98–108.

Polya, G. (2014). *How to solve it: A new aspect of mathematical method*. Princeton University Press.

Posavac, E. (2015). *Program evaluation: Methods and case studies*. Routledge.

Ravitch, D. (2001). *Left back: A century of battles over school reform*. Simon and Schuster.

Rittle-Johnson, B., Schneider, M., & Star, J. R. (2015). Not a one-way street: Bidirectional relations between procedural and conceptual knowledge of mathematics. *Educational Psychology Review, 27*(4), 587–597.

Rohlen, T. P., & Bjork, C. (1998). *Education and training in Japan* (Vol. 1). Taylor & Francis.

Romero, C., & Ventura, S. (2007). Educational data mining: A survey from 1995 to 2005. *Expert Systems with Applications, 33*(1), 135–146.

Roschelle, J., Bhanot, R. T., Patton, C., & Gallagher, L. (2015). *Strength of research for Reasoning Mind*. Menlo Park, CA: SRI International.

Roschelle, J., Gaudino, S., & Darling, S. (2016). Designing for consistent implementation of a 5th grade digital math curriculum. *International Journal of Designs for Learning, 7*(2), 41–56.

Russell, S. J., Norvig, P., Canny, J. F., Malik, J. M., & Edwards, D. D. (2003). *Artificial intelligence: A modern approach* (Vol. 2). Prentice Hall.

Schmidt, W., Houang, R., & Cogan, L. (2002). A coherent curriculum: the case of mathematics. *American Educator, (summer 2002)*, 1–17.

Schoenfeld, A. H. (2004). The math wars. *Educational Policy, 18*(1), 253–286.

Schwartz, D. L., & Martin, T. (2004). Inventing to prepare for future learning: The hidden efficiency of encouraging original student production in statistics instruction. *Cognition and Instruction, 22*(2), 129–184.

Shechtman, N., Roschelle, J., Feng, M., & Singleton, C. (2019). An Efficacy Study of a Digital Core Curriculum for Grade 5 Mathematics. *AERA Open, 5*(2), 2332858419850482.

Shute, V., & Ventura, M. (2013). *Stealth assessment: Measuring and supporting learning in video games*. MIT Press.

Slavin, R. E. (1990). *Cooperative learning: Theory, research, and practice*. Prentice-Hall.

Smith, S. M. (1984). A comparison of two techniques for reducing context-dependent forgetting. *Memory and Cognition, 12*(5), 477–482.

Stigler, J. W., & Hiebert, J. (2009). *The teaching gap: Best ideas from the 'world's teachers for improving education in the classroom*. Simon and Schuster.

Thomas, D. W., Burns, J., Audette, J., Carrol, A., Dow-Hygelund, C., & Hay, M. (2016). *Clinical development success rates 2006–2015*. Biomedtracker/Amplion.

Tucker, M. S. (2011). Standing on the shoulders of giants: An American agenda for education reform. Washington, DC: National Center on Education and the Economy.

Tuttle, C. C., Teh, B. R., Nichols-Barrer, I., Gill, B. P., & Gleason, P. (2010). Student characteristics and achievement in 22 KIPP middle schools: Final report. Mathematica Policy Research, Inc.

Valcarce, M. (2012). Russian mathematical pedagogy in Reasoning Mind. Retrieved from http://scholarship.claremont.edu/hmc_theses/30/.

VanLehn, K. (2011). The relative effectiveness of human tutoring, intelligent tutoring systems, and other tutoring systems. *Educational Psychologist, 46*(4), 197–221.

Veenman, M. V., & Alexander, P. (2011). Learning to self-monitor and self-regulate. In R. E. Mayer & P. A. Alexander (Eds.), *Handbook of research on learning and instruction*, 197–218. New York, NY: Routledge.

Walkington, C. A. (2013). Using adaptive learning technologies to personalize instruction to student interests: The impact of relevant contexts on performance and learning outcomes. *Journal of Educational Psychology, 105*(4), 932.

Waxman, H. C., & Houston, W. R. (2008). An evaluation of the 2006–2007 Reasoning Mind program. Retrieved from http://www.reasoningmind.org.

Waxman, H. C., & Houston, W. R. (2012). Evaluation of the 2010–2011 Reasoning Mind program in Beaumont ISD. Retrieved from www.reasoningmind.org.

Weber, W. A. (2003). An evaluation of the Reasoning Mind program at Hogg Middle School. Retrieved from www.reasoningmind.org.

Weber, W. A. (2006). An evaluation of the 2005–2006 Reasoning Mind project. Retrieved from www.reasoningmind.org.

Willingham, D. T. (2009). *Why 'don't students like school? A cognitive scientist answers questions about how the mind works and what it means for the classroom.* John Wiley & Sons.

Wu, H. (1999). Basic skills versus conceptual understanding: A bogus dichotomy in mathematics education. *American Educator,* 23(3), 14–19.

Wu, H. H. (2011). Phoenix rising: Bringing the Common Core state mathematics standards to life. *American Educator,* 35(3), 3–13.

About the Author

George A. Khachatryan is a cofounder of Reasoning Mind, a not-for-profit organization developing blended learning programs for elementary and middle schools. At Reasoning Mind, he oversaw the instructional design of many of the organization's programs, including *Foundations* and *Edifice*.

Dr. Khachatryan is a frequent speaker on education theory and practice and has been quoted by *Forbes*, *The Atlantic*, and other media outlets. Since 2016, he has been a member of McKinsey & Co., a global management consulting firm that serves leading businesses, governments, nongovernmental organizations, and not-for-profits. He continues to serve Reasoning Mind as a volunteer member of the board of advisors.

He studied pure mathematics at the University of Chicago, the University of Cambridge, and Cornell University, where he was a National Science Foundation graduate research fellow.

Index

Page references to tables and figures are indicated by *t*'s and *f*'s respectively.

For the benefit of digital users, indexed terms that span two pages (e.g., 52–53) may, on occasion, appear on only one of those pages.